When Machines Learn

AI and the Transformation of Society

by

Lars Meyer

When

Machines Learn

AI and the Transformation of Society

Table of Contents

Introduction

As the world pivots on the axis of technology, we stand at the brink of a new era, one underscored by rapid advancements in *Artificial Intelligence* (AI). The integration of AI into various spheres of our daily lives is no longer a distant prospect, but an unfolding reality. This book takes you on a deep dive into the complexities and wonders of AI, unraveling its layers to reveal a future interwoven with intelligent machines. Our journey will stretch from the nascent stages of AI's inception to its current state, and beyond, into the intriguing possibilities it holds for our collective future.

The advent of AI technologies has sparked a renaissance of innovation, bringing forth a surge of opportunities and challenges alike. At the heart of this renaissance lies the need to understand AI not just as a tool or a system, but as a transformational force poised to redefine what it means to live, work, and interact in a modern society. Our goal is to provide a comprehensive understanding of AI by highlighting its technological innovations, delving into the ethical quandaries it presents, and discussing its global implications. By doing so, we aim to prepare for the seamless integration and use of AI across diverse areas of life.

The overarching narrative of this book is speckled with testimonials of AI's prowess, cautionary tales, and the inspirational undertones of human resilience and innovation. We approach AI from multiple vantage points – dissecting its technical intricacies while also pondering over the societal ripples it creates. Each chapter of this book has been meticulously crafted to build upon the knowledge of AI, without overwhelming with esoteric jargon or technical overloads.

Within these pages, we encounter the fundamental constructs of AI, track its evolution through the annals of history, and spotlight the myriad applications that exemplify its potential. We ponder, with careful thought, the implications AI has for the workforce, including the disruption and creation of job markets, as well as the reshaping of professional skill sets and economies.

The ethical landscape of AI demands attention, and deservedly so. We shall engage with the moral framework that must be erected around the deployment of AI, analyzing the impact of biases in algorithmic decision-making and stressing the significance of transparency and accountability within AI systems.

Societal integration of AI stirs a myriad of reactions, from acceptance to resistance. Our investigation endeavors to address the mechanisms that can facilitate inclusive AI benefits and efforts to bridge any existing digital divides.

The sanctity of data protection and privacy emerges as a pivotal concern in the age of AI. Innovation must align with the protection of privacy rights, prompting a reassessment of data governance models and the role of technologies like encryption and anonymization in bolstering security.

Collaboration between humans and AI paves the way for augmented human capabilities and prompts a deeper inquiry into how such partnerships should be ethically designed for the benefit of society at large. Moreover, we consider the long-term implications of this symbiosis.

As a catalyst for innovation, AI has spurred a multitude of case studies proving its depth and agility. We delve into how enterprises and startups are collaborating with AI, and how intellectual property rights shape this landscape. Additionally, we examine AI from an international perspective, discussing how policies, regulations, and global trends are sculpting the terrain of AI development.

On a more intimate scale, AI affects our daily activities and choices. From fostering personalized experiences to redefining healthcare, education, and transportation, AI's touch on the human-machine relationship is evolving, necessitating a nuanced understanding.

Preparing for a future imbued with AI is crucial. This book lays out the educational reforms and anticipatory strategies that governments and individuals might employ to stay ahead of the curve.

As we draw conclusions, we recap the key technological breakthroughs, societal shifts, and lessons drawn from past experiences in order to better comprehend the present circumstances. Finally, we cast our gaze forward, speculating on the long-term impact and role of AI in our world, and chartering a sustainable path into this brave new world of intelligent machines.

With every chapter offering a different facet of AI's crystal, this book stands as a testament to human curiosity and our relentless pursuit of knowledge. Whether you're a seasoned technologist, a policy-maker, a student, or simply an inquiring mind, these pages beckon with insights and motivations, preparing you to face a world where AI is not an option, but a ubiquitous reality.

Through the compendium of our findings and deliberations, it becomes evident that we are not merely passive observers of this AI revolution but active participants shaping its course. Humanity's collective wisdom, ethical considerations, and innovative spirit are the true arbiters of how AI will mold our future. As we peel back the curtain on AI's global narrative, let's embark on this intellectual odyssey with an open mind, readiness for change, and an unyielding spirit of discovery.

Chapter 1:
The Evolution of Artificial Intelligence

From the cogs and gears of ancient automata to the sophisticated algorithms shaping our world today, the journey of artificial intelligence has been a tapestry of human ingenuity and relentless pursuit of the future. The evolution of AI is not just a timeline of technological milestones; it's a reflection of our collective dreams and the complex challenges we've overcome. As we delve into the realm of machine learning, deep neural networks, and AI applications that often outperform human capabilities, it's crucial to grasp the roots of these innovations, to appreciate the toil and triumphs that led us here. In exploring the early theoretical groundwork laid by visionary scientists and the subsequent crescendo of advancements, we not only honor the past but also enhance our understanding of AI's transformative role in our lives. You'll find that as AI gallops forward, becoming an intrinsic part of everything from healthcare to entertainment, it invites us to reimagine our relationship with technology—and with each other—in a constantly evolving dialogue between the creations we marvel at and the creators we are destined to remain.

Fundamentals of AI As we journey further into the realm of artificial intelligence, it's pivotal to establish an understanding of its core principles. Artificial intelligence, at its most basic, is the development of computer systems that can perform tasks typically requiring human intelligence. These tasks involve understanding natural language, recognizing patterns, solving complex problems, and even displaying elements of creativity.

The foundation of AI spans several fields, from computer science to cognitive psychology. The systems that AI encompasses seek to mimic or replicate human cognitive functions, learning and adapting from the data they process. What makes AI extraordinary is its ability to not just execute predefined commands or tasks, but to learn from prior experience and improve upon its subsequent actions.

Machine learning, a subset of AI, is where much of the magic happens. It's a method for data analysis that automates analytical model building. Through algorithms, computers can learn from data, identify patterns, and make decisions with minimal human intervention. It's the lifeblood of AI's effectiveness and is continually pushing the boundaries of what machines can do.

Deep learning, a further iteration within machine learning, uses neural networks with many layers (hence "deep") to analyze higher-level data. Deep learning allows machines to solve more complex problems and recognize patterns in a way that's reminiscent of human thought processes.

The integration of AI into various applications relies on this powerful blend of machine learning and deep learning. For instance, in image and speech recognition software, AI systems analyze vast quantities of data to learn how to recognize patterns and nuances in ways very similar to how humans do.

Another critical aspect of AI is its ability to process and analyze big data at speeds and accuracies that are beyond human capabilities. The velocity of data creation in today's world is exponential, and it's AI's inherent gift to be able to navigate this labyrinth of information, extracting valuable insights that can inform decisions and strategies.

To enable these complex functionalities, AI systems are built upon sophisticated models like decision trees, support vector machines, and ensemble methods. These models provide the frameworks for AI to engage in predictive analytics, natural language processing, and a range of other sophisticated tasks that transform how we interact with the world.

There are two general kinds of AI recognized in the field: narrow or weak AI and general or strong AI. Narrow AI, which is prevalent today, excels in performing specific tasks within a limited context, often outperforming humans in accuracy and speed. General AI, on the other hand, would have the capacity to understand, learn, and apply intelligence across a broad range of tasks, much like a human can.

Human oversight remains a vital component in the current landscape of AI. Although systems are becoming increasingly autonomous, the moral responsibility and final decision-making power remain with us. Deterministic algorithms, those that follow a precise set of rules, still underpin many AI systems, ensuring a degree of predictability and control over these rapidly evolving technologies.

With all these capabilities, it's crucial to remain cognizant of the hardware that supports AI's vast computations. Advances in processing power and data storage capabilities, epitomized by the development of graphics processing units (GPUs) and tensor processing units (TPUs), have significantly propelled the field forward. These technological advancements enable the intricate calculations and data storage necessary for AI algorithms to function effectively.

Lastly, the interdisciplinary character of AI cannot be overstated. A synergy of expertise from diverse fields – from neuroscientists studying the mechanisms of natural intelligence to ethicists examining the implications of non-human decision makers – is required for the responsible and effective development of AI. This convergence is what drives innovation and ensures that AI systems are designed with a holistic understanding of their potential impact.

Understanding these fundamentals forms a crucial substrate upon which further exploration of AI's application, integration, and implications can be built. As AI continues to evolve and integrate itself into the tapestry of our daily lives, grasping its foundational elements will be key to unlocking its full potential and ensuring it serves society in the most beneficial ways possible.

It is our responsibility as the stewards of this new era to ensure that we maintain a keen awareness of AI's capabilities and limitations. By fostering an intimate understanding of AI's bedrock principles, we can better navigate the ethical landscapes, societal challenges, and technological innovations that lie ahead.

In the subsequent chapters, we will delve into the historical development of AI, its burgeoning role in the workforce, ethical considerations, integrations into society, and much more. But it begins here: with a solid grounding in the fundamentals of artificial intelligence, we are better prepared to appreciate the fabric of opportunities and challenges that it weaves.

Historical Overview and Development The path of artificial intelligence has been a myriad of milestones and groundbreaking discoveries, each laying a stepping stone towards the complex and advanced AI systems we see today. Understanding the evolution of AI is akin to following a meandering river through time, which has expanded and deepened with each decade's technological advances.

The genesis of artificial intelligence as a formal field can be traced back to the mid-20th century when the term itself was coined at the famed Dartmouth Conference in 1956. At this seminal event, scientists like John McCarthy and Marvin Minsky set forth the ambitious goal of creating machines that could simulate every aspect of human intelligence. Such a goal seemed audaciously futuristic at the time, yet it has continued to inspire generations of researchers.

In the early days following the Dartmouth Conference, there was a surge of optimism. Researchers made bold predictions about machines that could translate languages, solve algebra word problems, and improve themselves. By the 1960s, algorithms like the Least Mean

Squares (developed by Widrow and Hoff) and the first neural network machine, the Perceptron (invented by Frank Rosenblatt), had emerged, displaying the capacity for AI to learn and adapt.

However, the path wasn't without setbacks. The 1970s brought a period known as the "AI winter," a time when the overly optimistic predictions of the 1960s were not met, and funding for artificial intelligence research dried up. Yet, even in these colder moments, the embers of AI did not extinguish. Researchers continued diligently on their quest, albeit with more realistic expectations. Expert systems like MYCIN, developed at Stanford, demonstrated that AI could make a significant impact, especially in specialized domains such as medical diagnosis.

The 1980s saw a reinvigoration in the field, bolstered by the advent of more powerful computers and sophisticated algorithms. The concept of machine learning, where computers could improve their performance through experience, began to redirect the focus of AI from purely rule-based systems to ones that could adapt and learn. It was during this decade that AI began to showcase its potential in industries like finance with algorithmic trading, and in manufacturing with robotics and automation.

Scientists and engineers continued innovating, integrating neural networks and Deep Learning by the 1990s and into the 2000s, which led to significant improvements in tasks like speech recognition and image identification. The game of chess, which had long stood as a measuring stick for cognitive prowess, was forever changed when IBM's Deep Blue became the first computer to defeat a reigning world chess champion, Gary Kasparov, in 1997.

Entering the 21st century, data became the lifeblood of AI, with the era of Big Data ushering in new opportunities. The concept of harnessing large datasets to train algorithms enhanced AI's capabilities, particularly with the introduction of the General Adversarial Network (GAN) in 2014, which could produce highly realistic data-derived outputs. This set the stage for AI systems not just to learn but to create.

The advent of smart personal assistants, such as Siri, introduced in 2011, and Alexa, introduced in 2014, marked a significant milestone in AI becoming an everyday convenience. AI was no longer a secluded research venture but a very real and tangible part of daily human life. Autonomous vehicles started to become a practical reality, and the dream of AI augmenting human capabilities took form in myriad ways, from warehouse robots to predictive text features in smartphones.

In parallel, the landscape of AI ethics began to form its foundation. With AI having a palpable impact on societies worldwide, topics of bias, transparency, and the impact of automation on the workforce came to the forefront. This dialogue became an essential part of the AI narrative, shaping its development towards a future that prioritizes inclusivity and fairness.

Today, AI's development continues at a breakneck pace, with advancements in machine learning algorithms, quantum computing, and natural language processing, just to skim the surface. The field stands on the cusp of new horizons, like AI's role in combating climate change or in conducting complex scientific research at speeds unimaginable to the human mind. The torches of innovation are being carried by corporations like Google, with its AI breakthrough AlphaGo, and OpenAI, with its versatile GPT models, among countless others.

As we piece together this historical tapestry, it is clear that the development of artificial intelligence bears the hallmark of human aspiration and ingenuity. Each chapter of its history is both a testament to past ambitions and a prologue to future achievements. The threads of philosophical musings, mathematical breakthroughs, technological advancements, and ethical considerations are woven tightly into the journey of AI.

With this historical context, we are better poised to appreciate the current state of artificial intelligence and the myriad of applications it has found in modern society. It's essential to recognize that AI's advancement is not just a chronicle of technological feats but also a canvas reflecting our collective societal values and choices.

The development of AI stands as a beacon of human progress, a field that continuously blurs the lines between what is possible and what resides in the hallways of our imaginations. It is our duty and privilege to navigate this journey responsibly, embracing the opportunities our forebears envisioned while forging pathways that honor our shared future.

As we delve further into the intricate interplay between humans and technology, let us carry the lessons from AI's history in our quest to develop a technology that elevates humanity. Just as each stride in AI has been marked by creativity and collaboration, so too will be our approach to understanding and influencing the continued evolution of this transformative domain.

Current Status and Exemplary Applications The present state of artificial intelligence represents a culmination of advancements bound by decades of research, development, and implementation. The current status is one of remarkable capabilities interwoven into the fabric of daily living, while also pushing the envelope of what's achievable through technology. In this era, AI systems are no longer mere novelties; they have become powerful tools that augment human abilities and catalyze innovation across myriad domains.

One of the most transformative applications of AI is observed in healthcare. Advanced algorithms assist in diagnosing diseases with precision rivaling seasoned professionals. Notably, AI's adeptness at sifting through vast datasets facilitates the early detection of conditions such as cancer, long before traditional methods might yield a diagnosis. This capability isn't just science fiction—it's currently saving lives and improving the efficacy and personalization of patient care.

Similarly transformative is AI's role in the automotive industry. Self-driving cars are increasingly becoming a reality, heralded by the gradual integration of AI features in commercial vehicles. These smart systems process data from sensors in real time to navigate traffic, detect obstacles, and reduce accidents—a promise of a future where road safety is substantially improved.

In the realm of customer service, chatbots and virtual assistants have become the norm. They not only respond to queries with human-like understanding but also learn from interactions to continually refine their responses and services. Companies are harnessing these AI solutions to enhance customer satisfaction and streamline operations. What was once a lagging area in business is now a thriving ecosystem of automated and intelligent interfaces.

Financial services have also been revolutionized by AI's analytical prowess, with algorithms that can detect patterns indicative of fraud. Banking institutions now utilize these systems to secure transactions and protect clients from sophisticated cyber threats. Investment firms employ AI to analyze market trends and generate predictions with stunning accuracy, shaping financial strategies in ways unattainable a few short years ago.

In education, AI's adaptive learning systems offer personalized learning experiences, catering to the specific needs and learning styles of individual students. These platforms can adjust difficulty levels, offer tailored resources, and provide instant feedback—an instrumental step in the evolution of education that echoes a student-centered approach.

AI's influence extends into the energy sector, optimizing grid operations through predictive maintenance and demand forecasting. By leveraging machine learning, energy providers can anticipate consumption spikes and redistribute resources accordingly, thereby reducing waste and enhancing efficiency in power distribution.

E-commerce platforms are using AI to create a more curated shopping experience. Recommender systems analyze previous purchases, search histories, and user ratings to suggest products that customers are more likely to buy. This not only helps businesses increase sales but also significantly enriches the customer experience.

The field of entertainment has experienced a surge of AI-driven personalization. Streaming services use sophisticated recommendation algorithms to suggest music, movies, and TV shows tailored to the individual tastes of their subscribers, effectively reshaping how content is distributed and consumed.

Smart homes and IoT devices, powered by AI, offer convenience and enhanced security by learning and adapting to residents' preferences and routines. From intelligent thermostats to voice-activated appliances, these AI applications promote energy efficiency and offer a glimpse into the potential of a fully interconnected home environment.

In agriculture, AI facilitates precision farming, which increases yield while minimizing environmental impact. Drones equipped with AI can assess crop health, monitor soil conditions, and administer treatments, ensuring resources are utilized with surgical precision.

Document analysis and natural language processing have reached unprecedented efficiency with AI, sorting, categorizing, and extracting valuable information from piles of unstructured data. This application is crucial for the legal and administrative sectors where document handling is pervasive and often cumbersome.

Even in creative fields, AI is making an impact. Generative AI can produce art, music, and literature that resonate with human experiences, sparking discussions about the nature of creativity and the role of machines in artistic expression.

Robotics, too, has been irrevocably altered by AI. With enhanced perception and decision-making capabilities, robots are performing tasks ranging from mundane assembly line work to complex surgical procedures. This union of robotics and AI is not just found in industrial settings; it's becoming part of everyday life.

The examples aforementioned merely scratch the surface of the current applications of AI. With a pervasive presence in nearly all industries, AI's current status is one not just of vibrancy but also of essentiality. Companies and institutions now view AI not as a luxury but as a key component of staying competitive and relevant in an increasingly digitized world.

For many, the rapid growth and integration of AI into our society evoke wonder and inspire innovation. AI's current trajectory suggests a future replete with opportunities and challenges alike—a dynamic that calls for thoughtful preparation and responsible stewardship. As new applications emerge and existing ones mature, our collective journey with AI continues to unfold in ways that promise to redefine the human experience.

Chapter 2:
AI and Its Impact on the Workforce

As we continue, it's crucial to understand how the unparalleled ascent of AI technology is poised to reshape the labor landscape—a transformation promising both upheaval and opportunity. The integration of ever-smarter algorithms into the fabric of our workday not only charts a new course for employment but engenders a profound shift in the very nature of work itself. AI's impact heralds a renaissance in productivity and innovation, yet simultaneously gives rise to legitimate trepidation over job displacement. The workforce is at a crossroads, one where routine, automatable tasks may dwindle, leaving room for ingenuity and high-skill endeavors to flourish. However, this is not a zero-sum game. The pervasiveness of AI will likely nudge industries toward embracing continuous learning and adaptation, thereby fostering the creation of new employment roles and the reskilling of the workforce. In navigating this seismic change, we must recognize that AI's emergence isn't a harbinger of obsolescence for human talent but rather a clarion call to harmonize our abilities with the inexorable march of technology—a call demanding both vision and action to guarantee future generations a place alongside their digital counterparts.

Emerging Trends in Employment As the landscape of artificial intelligence evolves, it is reshaping the way we think about work and employment. Pioneers in AI are forging new frontiers that promise to enhance productivity and catalyze growth across various sectors. However, with innovation comes transformation, and today's workforce faces a shifting paradigm that necessitates a closer examination.

The infusion of AI into the labor market is not a distant future scenario but a present reality, heralding both challenges and opportunities. Cutting-edge algorithms and automation technologies are increasingly performing tasks that were once firmly within the human domain. This trend leads to a multifaceted phenomenon that includes job displacement in certain sectors, as well as the creation of novel roles that demand distinct sets of skills.

One of the most notable trends in employment is the rise of the gig economy, facilitated by AI-powered platforms. These digital marketplaces offer unparalleled flexibility and have democratized the ability to earn, with AI streamlining the matching of gigs to workers. This transformation, however, raises questions about job security and benefits traditionally provided by steady employment.

There's also a progressive shift towards remote work. AI and related technologies are enabling a more distributed workforce, unbound by geographical constraints. While telecommuting is not new, AI enhances this capability by automating many collaborative tasks, making virtual teams more efficient and effective than ever before.

The workforce is witnessing the emergence of roles that focus on the oversight and improvement of AI systems. These 'AI Trainers' are professionals who teach AI applications how to perform their duties better. They curate data, refine algorithms, and ensure that AI systems operate within ethical guidelines.

Moreover, AI is accelerating the need for continuous learning and skill acquisition. As machine learning algorithms are applied to more complex tasks, human workers must adapt by mastering new technologies and methodologies. Lifelong learning is transitioning from a buzzword to a vital component of a sustainable career.

An undercurrent of the contemporary employment trend is the necessity for robust interdisciplinary knowledge. Careers are evolving to require a blend of skills that span technology, arts, and business. Workers need to be agile, with an ability to understand both the potential and limitations of AI within their domain.

Another trend to observe is the increase in demand for data scientists and engineers. Businesses are hungry for professionals who can interpret the vast seas of data generated by AI and extract actionable insights. These roles have become fundamental to decision-making processes in organizations striving to stay competitive in an AI-centric world.

AI is also fostering entrepreneurship by lowering barriers to entry. Startups can now leverage scalable AI solutions without the need for extensive resources. This democratization enables a surge in innovation, giving rise to new services and products that once seemed unfeasible.

In the realm of manufacturing, AI-driven automation is enhancing precision and efficiency. The integration of AI in this sector is leading to the emergence of 'smart factories', where predictive maintenance and optimized production lines significantly reduce downtime and boost throughput. This transition, however, is reorienting the workforce towards maintenance of machines and systems engineering instead of manual labor.

The healthcare sector is experiencing the introduction of AI-driven diagnostics and personalized treatment plans. These advancements necessitate a workforce that is not only clinically proficient but also technologically adept, capable of collaborating with AI to provide better patient outcomes.

AI is streamlining recruitment processes as well, with intelligent systems capable of sourcing and screening candidates more effectively. This has significant implications for human resources professionals, who now need to focus more on strategic human capital development than on administrative tasks.

Financial services are being transformed by AI, with algorithms now capable of fraud detection, risk assessment, and customer service. This introduces a cadre of technologically savvy finance professionals who can navigate the complex interplay between financial products and AI tools.

Lastly, a critical trend that's emerging is the importance of ethical considerations in AI deployment. There's a growing need for roles focused on the governance and ethical use of AI. Ethicists and policy makers specialized in technology are finding their skills in higher demand to ensure that as AI integrates into employment, it does so in a way that aligns with societal norms and values.

As we look towards the horizon, the convergence of AI and employment is an unmistakable force, altering the traditional trajectories of careers and industries. The workforce of tomorrow will be defined not by routine and replicable tasks but by the ability to harness the complementarities of human ingenuity and artificial intelligence. This presents an invitation for current and future generations to explore the synergies between human potential and AI prowess, striving for advancement while upholding the dignity of work in an era of unprecedented technological progress.

Job Displacement Concerns have surfaced with the surge in artificial intelligence applications transforming the workforce, creating a landscape that is as promising as it is challenging. As we delve into the intricate dynamics of AI's role in reshaping employment, it becomes evident that the tide of machine learning and automation is bound to redefine what work means to human society.

The narrative of job displacement due to technological advancement is not new. The industrial revolutions of the past have taught us that innovation can simultaneously generate new types of employment while rendering certain skillsets obsolete. However, the speed and sophistication with which AI can perform tasks pose an unprecedented challenge to the labor market.

Various sectors, from manufacturing to services, find themselves at a crossroads where repetitive and rule-based tasks are increasingly being executed by algorithms and robots. This transition inevitably raises apprehension about job security and the risk of widening economic disparities.

Data reveals a more nuanced story, however. Automation driven by AI is not a simple binary of job creation and destruction. It also encompasses job transformation, where human roles evolve in response to technological companions. For instance, a bank teller's role today is different from what it was two decades ago, now focusing more on customer relationship management than routine transactions.

Nonetheless, the apprehensions about job displacement cannot be dismissed. The World Economic Forum predicts that by 2025, automation will displace millions of jobs, but it will also create millions more. The catch lies in the fact that the jobs created may not be in the same regions or industries as those lost, leading to significant social and economic friction.

An important aspect of this transition is the skill mismatch. The jobs of the future may require digital literacy, data analysis expertise, or the ability to work alongside AI systems. Without a concerted effort to retrain and upskill the workforce, the mismatch between available jobs and employable skills could exacerbate unemployment woes.

The concern for job displacement resonates differently across demographic lines. Historically disadvantaged groups may face additional barriers to accessing the education and training needed for these new AI-augmented roles, reinforcing existing inequalities.

Additionally, the psychological impact of job displacement is profound. Beyond economic strain, loss of employment can erode a sense of purpose and identity that is closely tied to one's job. Society must grapple with these ramifications, addressing not only the economic but also the emotional support systems needed in such transitions.

Regionally, the impact of AI-driven automation is likely to vary. Cities with a heavy reliance on industries susceptible to automation may fear the hollowing out of their economic core, whereas those that are hubs of innovation may see a flourish of new job growth.

Policy interventions will play a critical role in mitigating AI-induced job displacement. These may include social safety nets, public-private partnerships to facilitate worker retraining, and incentives for industries that foster job creation. Sustainable workforce development is paramount, requiring coherent and forward-looking strategies from leaders.

Decisive action from educational institutions is also vital. Curriculum reform to incorporate AI-centric skills, starting from foundational education through to higher learning, is necessary to prepare the next generation for a transformed employment landscape.

One cannot ignore the opportunity within this upheaval. The potential for entrepreneurship and innovation in the era of AI is immense. As routine tasks are automated, human creativity and strategy can be unleashed on problems that machines are not adept at solving, such as those requiring emotional intelligence, nuanced judgment, and complex societal understanding.

While AI will certainly displace jobs, it can also enhance the quality of work by eliminating monotony and increasing efficiency. This, in turn, has the potential to boost job satisfaction for workers who are able to transition into these new roles effectively.

To realize the positive aspects of AI and minimize the negative impacts of job displacement, cohesive cooperation between government, industry, educators, and workers is necessary. Integrating AI into society is as much about managing its disruptive effects as it is about harnessing its transformative potential.

In conclusion, the concerns about job displacement in the wake of AI's ascendance are justified and complex. This challenge is not insurmountable, but it does demand an intelligent strategy towards re-skilling, evolving education systems, and crafting policy frameworks that support a flexible and resilient workforce. Embracing this holistic approach will lead us toward a future where AI acts not as an agent of displacement, but as a catalyst for human innovation and enterprise.

New Opportunities for Skilled Labor As we delve deeper into the impact of AI on the workforce, it's imperative to shift our focus toward the silver lining that is often overshadowed by the concerns of job displacement. Artificial intelligence is not just a harbinger of change but also a creator of vast new opportunities for skilled labor. No industry is untouched by the tendrils of AI, from healthcare to finance, which translates to a blossoming of roles for those ready to embrace and augment new technologies.

The narrative that AI will lead to a universal joblessness is not only overwrought but also incomplete. It fails to capture the dynamic nature of job markets and the human capacity for adaptation and growth. The rise of AI has, in fact, catalyzed the creation of jobs that require a nuanced blend of technical prowess and human ingenuity. Data scientists, AI specialists, and machine learning engineers are just the tip of the iceberg when it comes to emerging careers.

However, the opportunities are not limited to these high-tech roles. There's an expanding demand for cross-functional skills where individuals can bridge the gap between the digital and the traditional. Workers skilled in AI applications relevant to their field—such as digital marketing professionals proficient in AI-powered analytics or healthcare providers versed in AI-assisted diagnostics—stand at a significant advantage.

These changes in the labor market demand a proactive response from the educational sector. Given the fast pace of AI evolution, lifelong learning has become more important than ever. Vocational training programs and universities are pivoting towards courses that offer deep dives into AI and its applicability in various sectors, ensuring that the workforce remains competitive and agile.

The need for skilled labor also encompasses the ethical dimension of AI. As societies grapple with the implications of algorithmic decision-making, there is a growing need for individuals who can navigate the moral landscape of technology. Roles such as AI ethicists and compliance officers are becoming just as crucial as the technical personnel behind AI development.

AI also demands a higher level of specialization in traditional jobs. Trades such as electricians and mechanics are evolving, with new specializations in robotics and intelligent systems diagnostics. These professions are not being replaced; they are being upgraded, requiring a deeper understanding of the complex integration between hardware and AI-driven software.

Entrepreneurship is another avenue through which skilled labor can capitalize on the advent of AI. Startups focused on AI applications are sprouting up globally, tapping into niche markets and offering innovative solutions. Skilled professionals have the chance to lead the way in developing AI tools tailored to specific industry needs, from agriculture to entertainment.

Customer service and support roles are similarly transforming. AI in the form of chatbots and virtual assistants are handling routine inquiries, which in turn elevates the human role to tackle more complex and unique customer issues. The ability to work alongside AI and utilize it to enhance personal service and customer relationship management is a skill set in high demand.

Art and Creativity is yet another sector where AI is weaving new opportunities. Artists, writers, and designers are exploring AI as a tool to push the boundaries of creativity. The AI is not the artist, but a sophisticated instrument in the hands of the artist, enabling the creation of work that might be inconceivable without it.

Project management is undergoing a transformation as well. With AI providing valuable data insights and risk assessment, project managers can make more informed decisions and optimize workflows. This has elevated the role of project managers to strategic thinkers who can interpret AI data and apply it effectively to real-world applications.

Looking at the manufacturing sector, skilled laborers who can work alongside collaborative robots (cobots) are in high demand. These cobots are designed to work in tandem with human workers, not replace them. The key lies in understanding the operations and maintenance of such machines, ensuring smooth collaboration between human skills and robotics.

Even the world of finance is seeing a shift with the introduction of AI. Financial advisors and analysts who apprehend the insights generated by AI can provide better strategy and services to their clients. Combining financial expertise with AI fluency allows these professionals to cut through the noise and offer data-driven advice.

In the field of cybersecurity, AI offers the tools needed to combat sophisticated threats, but skilled cybersecurity professionals are the ones who wield these tools effectively. Their expertise helps organizations to anticipate and navigate the complex landscape of digital threats, making them invaluable to any security team.

Finally, the public sector and non-profit organizations stand to gain from skilled workers who can implement AI in addressing societal challenges. From optimizing resource allocation to enhancing disaster response, the possibilities are vast for those equipped to harness AI's potential for the public good.

The cultivation of new skills and the commitment to continuous learning are key aspects of thriving in this AI-augmented job market. As we continue to explore these new frontiers, it's clear that AI is not the end of the skilled labor story; it is an inspiring new chapter full of possibilities and promises for those ready to explore and exploit these frontiers.

Chapter 3:
Ethical Considerations in Artificial Intelligence

As we've explored the unprecedented expansion of AI and its burgeoning impact on the labor market, it becomes imperative to delve into the subsequent chapter of this narrative: the ethical landscape that underpins artificial intelligence. Ethical considerations in AI are not merely additional talking points; they form the backbone of a technology that holds the potential to reshape our society. As artificial intelligence systems become increasingly autonomous and integral to decision-making processes, it's critical that we confront the moral quandaries that accompany these advancements. The biases embedded within algorithms, the necessity for transparency in AI mechanics, and the accountability of AI systems are among the pertinent issues that demand rigorous scrutiny. Maintaining a moral compass in the deployment of AI is about safeguarding the very fabric of fairness and justice in our society. It's about ensuring that as machines learn, they don't inherit our fallibilities but help us transcend them, ushering in an era where technology amplifies our ethical ideals, rather than undermines them. In this chapter, we won't just untangle the complexities of ethics in AI; we aim to instill a sense of stewardship for those at the helm of AI development and deployment, cultivating an understanding that will empower us to steer this formidable technology towards a horizon of benevolence and human flourishing.

The Moral Framework of AI Deployment With the rapid pace at which artificial intelligence has woven itself into the fabric of society, it's imperative to pause and reflect on the moral scaffolding that must support such a profound revolution. As we delve into the ethical constructs necessary for the conscientious deployment of AI, we find ourselves grappling with complex dilemmas that transcend technological intricacies, reaching into the realm of human values and societal structures.

Morality, a construct typically reserved for human decision-making, extends its jurisdiction into the virtual brains of AI systems. These systems are being entrusted with decisions that affect human livelihood, well-being, and societal balance. As such, a fragrance of human morality must permeate the algorithmic corridors of AI development to ensure not only efficiency but also equity and justice in their operations.

AI deployment can be likened to entrusting a new member with societal responsibilities – one whose potential for impact is as vast as it is uncertain. Thus, creating a moral blueprint for AI assumes critical importance, acting as a navigational guide to prevent the straying of AI into ethical quagmires. This blueprint directs developers and users alike towards responsible and humane applications of machine intelligence.

Autonomy forms the first pillar in our moral framework. Ethicists and technologists contend with the dilemma of granting AI systems autonomy while ensuring they function within predetermined ethical parameters. How shall we imbue these systems with the 'wisdom' to make decisions that align with our broader ethical expectations? It's a question of designing constraints within which autonomy can operate safely and beneficially.

Beneficence - the imperative to do good - and nonmaleficence, the responsibility to do no harm, are cornerstones of ethical conduct that must extend to AI. These principles assume a weightier significance when the actions of AI systems can influence outcomes in healthcare, criminal justice, and economic sectors, among others. It's not merely about programming an AI to perform a task effectively; it's about ensuring that the task is performed in a manner that maximizes benefit and minimizes harm to all stakeholders.

Justice, another key tenet, involves distributing the benefits and burdens of AI technology equitably across society. The specter of AI exacerbating existing inequalities is a pressing concern that calls for proactive measures to ensure fairness in the dissemination and effects of AI. One must carefully navigate the fine line where technological progress does not come at the cost of widening socio-economic rifts.

Respect for human dignity and rights is a beacon that must guide the deployment of AI. In a landscape dotted with instances of technology being misused, the preservation of human dignity through respectful use of AI is a beacon we must steadfastly navigate towards. When AI touches on aspects of personal freedom, privacy, and individuality, its application must be sensitive to the inviolable respect owed to every person.

Transparency as a concept is twofold in the context of AI. Firstly, it comprises the 'interpretability' or 'explainability' of AI systems. Stakeholders affected by AI decisions have a right to understand the rationale behind AI-driven outcomes, especially when these decisions bear significant consequences. Secondly, transparency extends to the development process itself, fostering trust and enabling informed public discourse on the place of AI in society.

Our moral compass must also encompass accountability. When AI systems act amiss, determining liability and ensuring remediation is a labyrinthine task. Hence, clear frameworks of accountability are critical in ensuring that mechanisms are in place to address any potential misuse or unintended consequences of AI, thereby facilitating trust and confidence in these systems.

Collaboration emerges as a pivotal principle in the moral framework of AI. The intertwining of AI with societal fabric requires a multitude of voices and perspectives in its crafting. This includes ethicists, scientists, policy-makers, and the public coming together to shape a technology that is by the people and for the people. In such cross-disciplinary synergy, we find the checks and balances necessary to mold ethical AI.

Honing the principle of foresight involves anticipating the potential trajectories AI might take and understanding the long-term implications of its deployment. The stewardship of AI development demands not only addressing the immediate impacts but also evaluating future consequences, ensuring sustainability and the avoidance of detrimental societal shifts.

Lastly, resilience, both in ethical frameworks and AI systems, is indispensable. As AI systems encounter novel scenarios and as societies evolve, the capacity of AI to adapt ethically, without compromising moral standards, will be paramount. The onus lies on developers and regulators to build resilience into the very DNA of AI, facilitating a robust yet adaptable moral compass.

In sum, the moral framework of AI deployment isn't merely a set of directives to follow but a canvas, respecting past human experiences, present societal norms, and future aspirations. This framework is an ever-evolving organism, reflecting the dynamic interplay between technology and human ethics. It serves as a testament to the dedication to upholding humanistic principles in the ever-surging tide of technological advancement. Through this commitment, AI can serve as a testament to humanity's highest ideals and greatest strengths.

As such, an inclusive dialogue that spans across cultures, disciplines, and communities is key to effectively weaving these moral principles into AI systems globally. Diverse input ensures that AI deployment considers a wide range of ethical perspectives, reinforcing global solidarity and respecting the nuances of varied human contexts.

Above all, the moral framework of AI deployment urges us to envisage and strive for a symphony of technology and human values. It's a delicate yet profoundly powerful balance that can realize a future where artificial intelligence acts as a force for good, an enabler of human potential, and a custodian of ethical integrity. At the intersection of technology and morality, lies the opportunity to redefine the essence of progress – not just in terms of capability but also in compassion, fairness, and a shared commitment to the betterment of all.

Bias and Fairness in Algorithmic Decision-Making … Artificial intelligence has advanced to a point where algorithms can make decisions that used to be the exclusive domain of humans. It's an impressive feat that inspires awe, but one that comes with undeniable ethical complexities concerning bias and fairness. As we explore algorithmic decision-making, it's crucial to recognize that the algorithms, despite their potential for impartiality, are often only as unbiased as the data they are fed and the humans who craft them.

The issue of bias in AI systems can arise at multiple stages—from the collection and selection of data to the way an algorithm is programmed to interpret that data. An algorithm that is trained on unrepresentative or prejudiced data sets may inadvertently perpetuate or even amplify existing biases. This can result in discriminatory practices, such as preferential lending terms, biased hiring decisions, or unequal law enforcement.

To address such injustices, fairness must become an intrinsic part of each phase in the development of AI systems. Fairness, however, is a multifaceted concept. It's not about treating everyone exactly the same, as that may lead to disparities. Rather, ensuring fairness involves acknowledging diverse needs and circumstances and creating solutions that work equitably for all stakeholders.

Several types of biases need to be understood and mitigated. Historical bias reflects long-standing inequities in society; measurement bias arises from the way data is captured; algorithmic bias occurs due to flawed processing, and emergence bias evolves as systems interact with real-world feedback. Recognizing these biases is the first step towards establishing mechanisms for accountability in algorithmic decision-making.

There are emerging best practices aimed at reducing bias. This includes diversifying training data, performing regular audits, and implementing algorithmic impact assessments. Such practices can help illuminate the decision-making process and flag instances of potential unfairness before they cause harm. However, these measures are not foolproof; they must be complemented by human oversight.

Human oversight comes not only in the form of designers and data scientists continually tweaking the systems but also through appropriate governance frameworks. We must ensure there's a human touch, emphasizing empathy and understanding, to catch nuanced biases that even the most sophisticated algorithms may miss.

In the quest for fair AI, transparency is paramount. It's not just about opening up algorithms for inspection but also about clarifying the values and trade-offs embedded within them. Such transparency can help build trust in AI systems, making it easier for society to accept algorithmic decisions as both fair and legitimate.

Equity necessitates the purposeful inclusion of various perspectives in the conversation on AI. Stakeholders from marginalized communities must have a seat at the table during the development and implementation of AI systems to ensure that their insights and concerns shape the trajectory of technological development.

Yet, establishing fairness in AI is a dynamic challenge, not a static one. Society's norms and values evolve, and AI systems must be designed to adapt alongside them. Constant vigilance is required to continually assess and refine AI to maintain its alignment with societal ethics.

Education plays a critical role in promoting fairness in AI. By enlightening current and future generations about the intricacies of bias, we empower them to both anticipate and identify issues of fairness in AI, fostering a culture of responsible innovation.

Moreover, international cooperation can propel the pursuit of fairness in AI. A globally inclusive approach, recognizing the spectrum of cultural values and norms, can help set standards that prevent bias from undermining AI's potential benefits across borders.

Measuring the success of efforts to combat AI biases is difficult, yet methodologies are emerging. From quantitative metrics that track representation in data sets to qualitative analyses of user experiences, these evaluation tools guide progress toward impartiality.

In practice, combating bias in algorithmic decision-making is an intervention that requires a holistic approach, one that spans technical, organizational, and societal strategies. It is only through this comprehensive approach that AI can truly deliver on its promise of fair and equitable decision-making.

As we press forward, embracing the marriage of human intelligence with artificial counterparts, the call to action is clear. We must imbue AI systems with the ethical standards and compassion at the heart of human decision-making. In doing so, we unveil an era where AI serves as a beacon of fairness, significantly enriching the fabric of society.

The journey towards bias-free AI may be challenging, yet it is a necessary embarkation. It is an aspirational endeavor that speaks to the core of who we are as human beings—a symphony of diverse voices harmonizing around shared values of equity, fairness, and justice. As we weave these values into the algorithmic tapestries of tomorrow, we not only enhance the tools at our disposal but also the inherent dignity of our global community.

Transparency and Accountability in AI Systems are central themes in ethical artificial intelligence discourse. They are the pillars that maintain public trust and ensure responsible stewardship of technology. Without transparency, it's nearly impossible for users and stakeholders to understand how AI systems make decisions or produce outcomes. Without accountability, there's no mechanism to address when these systems cause harm or operate counter to societal values. The embodiment of these concepts within AI systems is not only a technical challenge but a multifaceted ethical imperative.

Transparency in AI refers to the clarity and understandability of AI processes and decisions to users, developers, and other stakeholders. AI must not be a black box, inaccessible to scrutiny. Instead, transparency entails that AI systems are open to examination, both in terms of their operational processes and the data they use. This visibility allows those affected by AI decisions to comprehend the rationale behind them and to identify potential biases or errors in the system.

But transparency alone is not sufficient. It needs to be complemented by accountability—a clear assignment of responsibility for the effects of AI systems. When AI systems act in ways that are harmful or discriminatory, it must be possible to hold those who built, deployed, or managed these systems responsible. This requires a clear framework of laws, regulations, and industry standards to ensure that individuals and organizations can be held to account for the outcomes of their AI systems.

Developing comprehensible AI depends on creating models that can explain their actions and decisions in human-readable terms. The movement towards "explainable AI" (XAI) is gaining momentum with increased demand for systems that provide insights into their decision-making processes. The goal here is to balance the performance of AI systems with explainability, which is often a trade-off with more complex models like deep neural networks.

In the spirit of accountability, regulatory efforts are growing. Governments around the world are beginning to propose and implement regulations that require AI systems to be transparent and operators to be accountable. The European Union's General Data

Protection Regulation (GDPR), for example, provides citizens with the right to explanation when they are subject to automated decisions.

The implementation of transparent and accountable AI also demands a cultural shift in the tech industry. There needs to be a commitment to ethics across the board, from the executive level to the developers on the ground. This kind of change requires education and training, as well as the establishment of ethical guidelines that are in line with societal values and the public good.

Auditability is an essential feature of AI accountability, where independent third-parties assess AI systems for fairness, security, privacy, and robustness. Regular audits can provide reassurances to the public that the AI systems are working as intended and not causing unintended harm. These audits become crucial, particularly in high-stakes areas such as health care, law enforcement, and financial services.

The pursuit of responsible AI also includes the question of redress. When AI systems do cause harm, there needs to be a clear and accessible mechanism for those affected to seek redress. This could be through legal channels, industry-specific tribunals, or regulatory bodies. The entity responsible for the system's deployment must be prepared to take corrective action.

Consumer and end-user education plays a crucial role in transparency and accountability, too. If users are unaware of how AI systems impact their lives or they lack the knowledge to question these systems' decisions, they are at a significant disadvantage. Therefore, raising public awareness and understanding concerning AI is fundamental.

Transparency tools and frameworks are evolving alongside the AI systems themselves. AI developers are increasingly using tools such as model cards, data sheets for datasets, and impact assessments to communicate with stakeholders about how their AI systems work and the considerations that have been taken into account in their development.

Trust is the foundation upon which the relationship between AI and society is built. An AI system that is both transparent and accountable is more likely to be trusted. While achieving complete trust in AI may be a distant goal, the pursuit of these principles is essential for the healthy integration of AI into the daily fabric of society.

In addition to these measures, industry consortia and multistakeholder groups are emerging to tackle the challenges of AI governance. By working together, these groups can share best practices and develop frameworks for responsible AI that balance innovation with ethical considerations.

Finally, accountability does not end once an AI system is deployed. Post-deployment monitoring is vital to ensure that AI systems continue to perform ethically throughout their lifecycle. Continuous monitoring might catch issues with data drift, model decay, or changing societal values that necessitate alterations to the AI system.

With the stakes so high, a failure to address transparency and accountability could lead to repercussions that range from the erosion of civil liberties to unintended escalations of inequality. Attention to these areas is critical if AI is to fulfill its promise of being a force for good.

Diverse perspectives add rich layers to the conversation about AI transparency and accountability. By including voices from different industries, cultural backgrounds, legal frameworks, and lived experiences, the landscape of ethical AI can be robustly mapped out, highlighting common challenges and unique circumstances. This diverse input is essential for developing AI systems that serve everyone rather than a select few.

In essence, transparency and accountability are not just technical issues; they are societal imperatives that underpin trust in AI systems. As we venture further into a world augmented by artificial intelligence, the commitment to these principles must grow stronger, ensuring that AI serves humanity with fairness, reliability, and openness.

Chapter 4:
AI Integration in Society

As we transition from the ethical frameworks discussed in the previous chapter, it's crucial to explore the integration of AI within the fabric of our society—a complex tapestry of acceptance, resistance, and necessity. The seamless incorporation of artificial intelligence into our daily lives hinges on creating a consonance between innovative technology and the diverse human conditions across global communities. This chapter delves into the societal dimensions of AI deployment, examining how acceptance levels vary across different demographics and cultures. We'll consider strategies to ensure that the benefits of AI are inclusive, addressing the risk of widening the digital divide that could marginalize underrepresented groups. In understanding the social dynamics at play, we seek to build a coalition of human-centric AI applications that not only resonate with societal needs but also uphold our communal values, knit our communities closer together, and elevate our shared humanity.

Social Acceptance and Resistance As we navigate through the intricacies of Artificial Intelligence and its encroachment into various aspects of our society, we come face-to-face with a myriad of responses. The topic of social acceptance and resistance is twofold in its complexity. On the one hand, there are those who embrace AI for its potential to improve quality of life and drive human progress. On the other, resistance bubbles up from deep-seated fears and genuine concerns over its possible repercussions.

Ensconced in the wide chasm between these two extrema is a delicate interplay of human psychology, cultural expectations, and socio-economic factors. Acceptance of AI is often driven by perceived benefits such as increased efficiency, cost savings, and novel conveniences. In sectors such as healthcare, where AI can aid in diagnosis and patient care, the benefits are tangible and can lead to widespread support.

Yet, sometimes acceptance is gradual, shaped by exposure and familiarity. As people interact more with AI-enabled services and products—personal assistants, chatbots, or recommendation systems—they gradually adapt to and even come to expect these enhancements. This normalization of technology is a critical step toward broader social acceptance.

Resistance, however, cannot and should not be dismissed as a mere aversion to change. It often surfaces from genuine concerns over job security as automation becomes prevalent. People worry that their livelihoods could become obsolete, a topic carefully examined in previous chapters. The threat of unemployment due to AI-induced automation is a catalyst for resistance and calls for conscientious integration strategies.

Furthermore, ethical concerns play a pivotal role in societal resistance. Questions around bias, discrimination, and the invasion of privacy that can arise with AI systems foster skepticism. Issues of transparency and accountability, touched upon earlier, feed into a narrative that AI might operate without human-aligned moral compasses or responsibility, exacerbating resistance.

The cultural fabric of a society also influences the level of acceptance. Cultures that prioritize collective wellbeing might view AI differently from those that place higher value on individual liberty and privacy. This nuanced intercultural perspective on AI brings to light the fact that acceptance is not a one-size-fits-all notion; it demands localization of approach and understanding.

Educational disparities too frame the conversation around AI. Individuals with a deeper understanding of AI's capabilities and limitations may be more open to its integration. Conversely, a lack of understanding can breed fear, leading to resistance. It's therefore essential to educate the public on these issues, a point we'll expand on in later chapters on educational reforms.

Moreover, witnessing AI's impact on the world stage can shape public opinion. Stories of AI's involvement in unfair practices or even accidents involving autonomous vehicles can leave lasting impressions. These instances highlight the fact that even isolated incidents of failure or misuse can fuel widespread resistance and counter the efforts made toward acceptance.

Regulatory environments too play a crucial role. When governments show enthusiasm for AI and implement supportive policies, it signals a level of safety and reliability to the populace. In contrast, a lack of regulation or reports of government misuse of AI technology can breed distrust and resistance within society.

Economic benefits—while persuasive—are not the panacea for overcoming resistance. Even in scenarios where AI contributes to economic growth, if the wealth generated is not perceived to be distributed equitably, social resistance may persist. Thus, the narrative of AI's benefits must be inclusive and equitable to promote wider acceptance.

Attitudes toward AI are also invariably linked to the media's portrayal of the technology. Sensational headlines and dystopian tales in fiction can unduly influence the public's perception. To cultivate a balanced view, it's imperative that accurate information and success stories of AI are also at the forefront of media coverage.

Engagement and participation of the public in AI development processes can further alleviate fears. When people feel that they have a say in how AI is being integrated into their lives, they tend to show more acceptance. As a society, including diverse voices at the AI discussion table is not just desirable, but necessary for harmonious integration.

As we move forward, the dynamics of social acceptance and resistance will continue to evolve. AI is not simply a technological revolution; it is a societal transformation that requires empathy, dialogue, and thoughtful action. Advocacy and open channels of communication stand as the bedrock for navigating the waters of resistance and planting the seeds of acceptance.

Ultimately, the interplay between acceptance and resistance is a tale of human engagement with innovation. As with any major technological advancement in history, AI presents both challenges and opportunities. Harnessing the opportunities while judiciously addressing the challenges will determine the trajectory of AI's acceptance in society. The following chapters will further delve into how we can facilitate inclusive AI benefits and bridge the digital divide, thereby shaping a future where AI becomes a trusted and integral part of our societal fabric.

Facilitating Inclusive AI Benefits

As our exploration of AI's role in society deepens, a defining challenge is ensuring that the advantages of AI do not become a privilege for a select few, but a shared benefit for all. Inclusive AI is paramount not only to prevent exacerbating existing inequalities but to actively reduce them. This entails a multifaceted approach starting from the inception of AI technologies to their deployment and beyond.

Firstly, inclusivity requires that AI systems be designed with diversity in mind. It's not enough for these systems to function effectively; they must do so in a way that serves diverse populations equitably. If they're trained on datasets that do not reflect the full spectrum of humanity, biases will inevitably skew their functionality, resulting in unequal service and opportunities.

Secondly, when we consider the development and maintenance of AI systems, we must encourage participation from individuals across all strata of society. A more diverse group of creators and maintainers ensures a wider range of issues are foreseen and addressed before products reach the public. The opportunity to contribute also empowers communities and drives innovation from varied perspectives, fostering better AI solutions.

It's also crucial to prioritize education and awareness among all sections of society. Knowledge is power, and simplifying the basic tenets of AI so they're accessible to everyone means that more individuals can understand, engage with, and potentially shape the AI conversation. Cultivating a populace that is AI-savvy is a stepping stone toward an inclusive AI ecosystem.

Moreover, meaningful legislation is key. Governments need to enact policies that incentivize the creation and adoption of inclusive AI. This includes setting standards for data collection and AI training processes which promote diversity and equity, as well as regulatory frameworks that hold companies accountable for the social impact of their AI systems.

Investment in infrastructure is another pillar to enabling inclusive AI benefits. Without adequate access to the internet and digital resources, many will be barred from the benefits that AI can deliver. Expanding digital infrastructure to underserved and rural communities is essential to closing this access gap.

In a closely related vein, private and public sectors must collaborate to drive the affordability of AI-driven services and products. If AI technology remains out of financial reach for many, its benefits will remain concentrated among those who can afford it. Here, subsidies, scaled pricing, and innovative financial solutions may play a role.

When it comes to employment and AI, it's crucial to create pathways for individuals whose jobs are most at risk of being supplanted by AI. Reskilling and upskilling programs should be accessible and affordable to provide these workers new opportunities in the emerging economy. Organizations, too, can contribute by investing in their employees' continuous learning and adaptation.

Access to healthcare is another area where AI has the potential to make a profound impact. Telemedicine and AI-powered diagnostic tools can reach underserved communities, but only if the underlying technology accounts for and adapts to the varied environments and genetic diversities it will encounter in different populations.

In the realm of education, AI could democratize learning and tailor experiences to individual needs, but this only happens if all students have equal access to these technologies. Efforts need to be made to equip schools across varying socio-economic locales with the tools and connectivity they need to integrate AI into their learning environments.

There are also considerations regarding ethical AI governance. Who is at the table when decisions are made about AI development, usage, and regulation matters. Ensuring diverse representation within these bodies counters narrow viewpoints and vested interests, promoting decisions that consider wide-ranging social implications.

Public awareness and engagement campaigns are also significant. They not only educate individuals on the benefits and risks associated with AI but also invite public discourse and participation. Ensuring that everyone's voice can be heard in conversations about AI is a stride toward shaping a technology that serves the collective good.

Transparency in AI processes and systems is yet another necessary factor for inclusivity. Users must understand how AI systems make decisions and what data they use. This is not to overwhelm with technicalities, but to foster trust and allow users from all backgrounds to engage with AI confidently and critically.

Continual monitoring and assessment keep us vigilant. As AI technologies are implemented, it's imperative to study their societal impacts, adjusting and improving systems where disparities are found. It's a process of continuous improvement and vigilance to uphold inclusive values.

In the end, the benefits of AI can't just serve the economic bottom line; they must resonate with the moral imperative of inclusivity. When AI systems are crafted with empathy, operate within a fair legislative framework, and are embraced by an informed, engaged, and empowered society, they have the potential to co-create a future where prosperity is accessible to all.

In summary, while the potential of AI is boundless, so too should be the access to its benefits. As we move forward, it's our collective responsibility to ensure that AI technologies are harnessed to not only revolutionize industries and economies, but to uplift every member of society, to create a shared and sustainable future enriched by intelligent, inclusive technology.

Bridging the Digital Divide Within the vast expanse of AI integration in society, a critical challenge that emerges is the digital divide—a term symbolizing the gap between demographics and regions that have access to modern information and communication technology and those that do not. This divide encompasses not only the accessibility of physical technologies but also the capability to use them effectively. In this section, we delve into a comprehensive exploration of how AI can be a tool to narrow this divide, designing a path towards greater digital equity.

Initiatives to overcome the digital divide are multifaceted, requiring a blend of public policy, education reform, and innovative technology solutions. The starting point often involves enhancing the infrastructure that allows greater connectivity. This includes expanding high-speed internet access to underserved communities, often in rural or impoverished regions. Advances in AI can streamline the deployment of such infrastructure, predicting the most impactful areas for high-speed internet installation using vast datasets encompassing geographical and socioeconomic factors.

Beyond connectivity, access to appropriate hardware is necessary for individuals to interact with AI-driven services. Subsidization of devices for lower-income households could democratize the benefits of the digital revolution. By implementing AI in manufacturing and supply chain logistics, these technologies can be made more affordable and thus more accessible. Furthermore, educational AI platforms have the potential to personalize learning experiences, making digital literacy more attainable for all levels of society.

Education plays a cornerstone role in closing the digital divide. Knowledge and comfort with digital tools are paramount for people to utilize AI's full potential. Educational programs targeting diverse age groups can develop competencies that allow for meaningful engagement with technology. AI can enhance these educational strategies through adaptive learning systems that cater to the individual's pace and style, providing a custom-fit education model reminiscent of private tutoring.

Local libraries and community centers have been traditional bastions for community education, and they also have a role to play here. These centers can become access points for high-speed internet and AI educational tools, offering workshops and resources to community members. By leveraging AI to personalize the learning experience, these community hubs can facilitate lifelong learning and digital proficiency.

Engagement with the private sector is crucial to bridging the gap. Partnerships with technology companies can provide the hardware and software necessary at lower costs or as part of corporate social responsibility initiatives. These collaborations can also lead to innovative approaches to technology distribution, such as implementing device lending programs or subsidized technology exchange platforms, ensuring that access to AI-powered devices isn't confined to the wealthy.

Representation matters in the realm of AI development. Including a diverse array of voices in AI design and decision-making ensures that the outputs of AI are well-suited for a broader population. Recruitment policies in the tech industry must therefore be geared toward diversity to tap into a pool of talent that reflects the global mosaic of AI end-users, contributing further to bridging the digital divide.

Up-to-date skills are necessary for the workforce of tomorrow. Reskilling and upskilling initiatives are imperative as AI transforms job markets, ensuring that workers are not left behind. Tailored AI training programs can chart pathways for career development, aligning workforce skills with the evolving needs of a digital economy and fostering a culture of continuous learning.

Governments have a pivotal role in making AI a public good rather than a source of further inequality. Policies aimed at providing equitable access to AI resources, improving digital literacy, and fostering innovation can set a foundation for digital inclusion. Governments can incentivize the development and dissemination of AI through grants and tax breaks, supporting projects that aim to reduce the digital divide.

Nonprofits and international organizations also have significant parts to play, acting as the bridges between governments, businesses, and underrepresented communities. By aligning efforts with the Sustainable Development Goals (SDGs), particularly SDG 9 (industry, innovation, and infrastructure) and SDG 10 (reduced inequalities), these organizations can guide AI's role in creating inclusive and resilient societies.

Privacy and security are major concerns for all but are especially important for vulnerable populations new to digital technologies. AI systems can be designed with robust security measures that protect against data breaches and misuse. Education on digital citizenship, which includes understanding digital rights, responsibilities, and security practices, is indispensable in any program aimed at bridging the digital divide.

Local initiatives should not be underrated in their power to catalyze change. Grassroots movements provide unique insights into the needs of specific populations and have the agility to adapt and iterate solutions rapidly. Pairing local knowledge with AI's analytical capabilities can result in targeted interventions that address the specific challenges faced by a community when accessing technology.

Public awareness campaigns can underscore the importance of technological inclusivity, creating a societal imperative to act. Through storytelling and case study dissemination, the successes of bridging efforts can inspire further action and investment in closing the digital divide. Highlighting the life-changing impacts that AI can have when it reaches across the digital chasm can galvanize public support and advocacy for digital inclusion initiatives.

International cooperation is paramount in addressing the global scale of the digital divide. Knowledge sharing, cross-border partnerships, and harmonized strategies can ensure that innovations in AI are leveraged to foster global digital equity. Global forums dedicated to AI and technological ethics can be instrumental venues for fostering this type of international dialogue and coordination.

Finally, monitoring and evaluation of AI's role in bridging the digital divide require persistent attention. Data-driven assessments that measure the impact of initiatives are vital for understanding what works and what doesn't. Adjustments and scaling up on effective programs can be made, perpetuating a cycle of continuous improvement aimed at eliminating the digital divide for good.

In examining the narrative of AI integration within society, bridging the digital divide stands out as a moral imperative. As we tread deeper into the AI-augmented era, the concerted efforts of governments, businesses, nonprofit entities, and communities can structure the landscape of technology to be inclusive, rich in opportunity, and a domain where equity is the norm, not the exception. It's this intersection of technology and humanity that will define how the story of AI unfolds in the lives of individuals across all spectrums of society.

Chapter 5:
Data Protection and Privacy Concerns

After recognizing the intricate web of societal implications in the AI landscape, we cannot overlook one of the most critical considerations: the safeguarding of personal data and privacy. In this era where data is heralded as the new oil, it fuels advancements in AI but also raises substantial concerns about its potential misuse. Citizens around the globe are rightfully anxious about how their sensitive information is handled, shared, and protected, igniting a pressing discourse on privacy rights. Robust encryption methods and anonymization techniques have become indispensable for fortifying data against breaches, yet they sometimes lock horns with the impetus to harness AI's full potential. As stewards of such powerful technologies, there's a responsibility to erect sustainable data governance frameworks that uphold the integrity of individual rights without stifling innovation. This chapter is a deep dive into the tumultuous waters of data protection, where we'll navigate through the storm of ethical quandaries and steer towards calmer seas where privacy is not just an afterthought but a cornerstone of the AI environment.

Balancing Innovation with Privacy Rights

As we delve into the intricacies of data protection and privacy concerns, a critical balance emerges—one that challenges innovators and policymakers alike. This balance pertains to the harmonization of technological advancement with the fundamental privacy rights of individuals. While AI promises an era of unprecedented growth and opportunities, it rides a fine line that could potentially infringe upon the privacy of its users.

The rapid evolution of AI technologies has brought forward tremendous capabilities to analyze, predict, and influence human behavior. These capabilities, however compelling, must be carefully weighed against the right of individuals to maintain control over their personal information. Societal values demand the preservation of personal autonomy, even amidst the whirlwind of digital transformation.

One of the fundamental aspects of this balance is consent. Innovators and AI system builders must ensure that they obtain informed consent from individuals when their data is collected, processed, or shared. Transparency in how data is used and for what purpose is crucial, as it empowers users to maintain a level of control and understanding over their digital footprint.

Privacy cannot be an afterthought in the age of AI; rather, it must be embedded within the design and deployment of new technologies. Privacy by Design—a concept that advocates for privacy to be incorporated into technology at the onset, rather than appended later—is more relevant than ever. This approach demands meticulous planning and a paradigm shift in how we conceptualize innovation.

Data minimization is another key principle, urging the collection and storage of only the data needed for a specified purpose. AI systems often have the ability to extract more information than necessary for their function, creating risks for excessive data harvesting that can be intrusive.

It's essential that legal frameworks keep pace with technological advancements to protect privacy rights effectively. As AI systems become more sophisticated, regulation must evolve to address the complexities of new data ecosystems. Enforcement is crucial, but it requires governments and international bodies to be proactive, rather than reactive, in their legislative efforts.

One of the more controversial aspects of AI innovation is its ability to leverage big data for profiling and predictive analytics. While these methods can unlock incredible possibilities for personalization and efficiency, they also raise concerns about the potential for surveillance and loss of anonymity. The ethical development of AI must therefore involve strict boundaries to prevent the misuse of personal profiles.

User empowerment is a valuable asset in the quest for equilibrium between privacy and innovation. Giving users the tools and knowledge to manage their privacy settings, understand data collection processes, and discern the intentions behind AI interactions is vital. This empowerment enhances public trust in AI technologies and encourages responsible usage.

Moreover, the role of education in this field cannot be understated. Awareness among the general population about their rights and the implications of AI on those rights is paramount. Educational initiatives should be aimed at demystifying the technology and elucidating the value of privacy in a digital world.

Anonymization and pseudonymization are techniques increasingly used in AI systems as a means to protect individual privacy. These tools, when applied effectively, can reduce the risks associated with personal data breaches and misuse. However, it's important to recognize that these methods are not foolproof, and continuous advancements in AI capabilities might challenge their reliability.

Another significant aspect in maintaining this balance is the concept of purpose limitation. AI systems must be designed with clear boundaries regarding the intended use of data. Any temptation to use data for secondary purposes that haven't been expressly agreed to by the data subjects should be strictly curtailed, ensuring fidelity to the original terms of data consent.

Encouraging ethical innovation requires the creation of environments where developers and firms are rewarded not only for groundbreaking work but also for responsible practices. Encouraging a culture of ethical technology development, where privacy protections are seen as a competitive advantage, can stimulate positive change industry-wide.

Finally, the dialogue between innovation and privacy rights is not a static one. As AI continues to evolve, so too must our understanding and protections of privacy. Continuous monitoring and assessment, including impact assessments for AI deployments, can ensure that the balance is maintained even as new technologies emerge.

It bears emphasizing that the solutions can't rest solely on the shoulders of technologists and regulators. A societal consensus on the value of privacy in the digital age, along with active engagement from individuals and advocacy groups, will cement the foundations that safeguard our privacy rights in the face of relentless innovation.

In conclusion, balancing innovation with privacy rights in the realm of AI is neither quick nor simple. It's a deliberate process, demanding collaboration, foresight, and a persistent dedication to individual rights and societal values. As we forge ahead, this balanced approach will form the cornerstone of ethical AI integration, promising both the advancement of technology and the protection of privacy in the digital age.

Data Governance Models As an essential part of the framework that enables the responsible use of artificial intelligence, data governance models are vital to ensuring that data is handled in a manner that protects privacy, maintains integrity, and sustains the trust of all stakeholders involved. Within this ecosystem, various models of data governance have emerged, each tailored to the requirements of different types of organizations and regulatory environments.

Data governance models are the blueprints for managing data assets, and they guide organizations on how to collect, store, manage, share, and protect data. These models are critical because they help to maintain data quality and data management standards across the entire lifecycle of data. Without a robust governance framework in place, the vast and complex datasets that feed AI systems can quickly become unmanageable and can lead to misinformed decisions, potential breaches, and erosion of user trust.

One of the foundational data governance models is the *Centralized Governance Model*. This approach is characterized by a single point of authority within an organization that defines and enforces data management standards. This model offers clarity and consistency in data handling but can sometimes be less agile in adapting to the rapid changes typical of AI-driven initiatives.

In contrast, the *Decentralized Governance Model* distributes data governance responsibilities across different organizational units. While this decentralized approach can enhance agility and cater to department-specific needs, it might lead to inconsistencies in data handling if not coordinated effectively.

Another influential model is the *Hybrid Governance Model*, which combines elements of both centralized and decentralized governance. It seeks to strike a balance between the two, offering standardization where necessary, while still allowing individual departments or business units the flexibility to address their unique data requirements.

Emerging from the need for granular control over data, is the *Data Stewardship Model*. In this approach, data stewards are appointed to oversee data assets at a detailed level. Stewards are typically experts in their data area, ensuring that data is used both responsibly and strategically to achieve organizational goals.

The *Federated Governance Model* leans towards a coordinated but separate governance structure, where different units across an organization align with a common set of standards and practices but retain autonomy. This collaborative approach often finds a place in large conglomerates or multinational corporations that must adhere to diverse legal and regulatory environments.

Looking beyond traditional organizational boundaries, the *Open Data Model* encourages transparency and accessibility. Public sector entities and international organizations often utilize this model to provide datasets that can be accessed by the public for research, development, and other purposes, thereby fostering innovation and civic engagement.

With regulatory compliance at its core, the *Regulatory Compliance Model* is focused on adherence to laws and regulations. Companies that operate in highly regulated industries, such as finance and healthcare, often adopt this model to ensure they are meeting legal obligations regarding the handling of sensitive data.

In the context of AI, a governance model that is gaining traction is the *Responsible AI Governance Model*. This framework emphasizes ethical considerations in the data lifecycle and the design of AI systems, advocating for fairness, accountability, and transparency. It calls for embedding ethical principles into every stage of the AI development process, from data collection to model deployment.

Environmentally conscious organizations are beginning to look toward the *Sustainable Data Governance Model*. This model ties in data practices with environmental sustainability, advocating for efficient data storage, processing, and disposal practices that minimize carbon footprints and promote the green use of data-related resources.

Furthermore, the rise in the volume and variety of data has paved the way for the *Big Data Governance Model*, which is tailored to handle the complexities associated with large scale data sets. This model emphasizes the importance of data quality, lineage, and the integration of advanced analytics to glean valuable insights from vast datasets.

Each of these data governance models must grapple with the often competing interests of innovation and risk management. The goal is to enable organizations to leverage their data effectively for AI initiatives while ensuring that all precautions are taken to protect the data and respect user privacy.

Interconnectivity in the modern world has also led to the development of the *Cross-Organizational Governance Model*, where data governance extends beyond the boundaries of a single entity. This collaborative framework involves multiple organizations agreeing on shared governance standards, crucial for AI projects that rely on aggregated or pooled datasets.

Implementing an effective data governance model requires a tailored strategy that factors in organizational culture, technology infrastructure, data types, regulatory requirements, and the specific goals of the AI systems to be deployed. It's a dynamic and ongoing process, designed to flex and evolve along with changes in the AI landscape, ensuring relevance and efficacy.

Ultimately, the right data governance model enables organizations to harness the full potential of artificial intelligence securely and ethically. The journey towards integrating AI into various aspects of life demands that we handle the underlying data with the utmost sensitivity and foresight, afforded by a clear, structured, and adaptive governance framework.

The convergence of robust data governance with advanced AI technology can arm organizations with the tools necessary to not only innovate but also cultivate trust, forge a competitive edge, and contribute to the responsible growth and governance of AI. As we progress deeper into the age of artificial intelligence, embracing and refining these data governance models becomes not just a strategic move, but a foundational element of sustainable and ethical AI deployment.

The Role of Encryption and Anonymization As we delve further into the intricacies of data protection and privacy concerns within the realm of Artificial Intelligence (AI), it is crucial to critically examine the role of encryption and anonymization. These two processes are fundamental to safeguarding sensitive information, ensuring that as AI systems process vast datasets, individual privacy is not compromised.

Data is the lifeblood of modern AI systems. It powers machine learning algorithms, enabling them to make accurate predictions and carry out tasks with a degree of intelligence that mirrors, and in some instances, surpasses human capabilities. However, this reliance on data posits a significant challenge: How can we maintain the sanctity of personal information while still reaping the benefits of data-driven AI?

Encryption serves as one of the bulwarks against unauthorized access to personal data. By transforming information into a format that is unreadable without a specific key or cipher, encryption ensures that data intercepted by unintended parties remains safeguarded from exploitation. In the context of AI, where data interchange and storage happen at unprecedented scales, encryption is not merely an option but a necessity.

The integrity of encryption relies on sophisticated algorithms. The complexity of these algorithms must stay apace with computational advancements to thwart efforts by malicious actors who aim to decipher private information. As quantum computing looms on the horizon, promising to revolutionize computational capacities, encryption methods need to evolve to withstand potentially groundbreaking decryption techniques.

Anonymization complements encryption by stripping away identifiable information from data sets, thus ensuring individuals cannot be traced or identified from the data. This process is incredibly pertinent when AI systems are trained on datasets containing personal attributes. Anonymization allows for the utility of data while negating the potential for privacy intrusion.

Differential privacy is an emerging form of anonymization that incorporates randomness in data processing. By adjusting the information ever so slightly, differential privacy ensures the outcome of any analysis is not significantly altered if any one person's information were removed. This furnishes an additional layer of protection, affording peace of mind to individuals whose data is part of an AI system's learning material.

Yet, encryption and anonymization are not without their challenges. The quest for balance between utility and privacy entails careful consideration. If data is overly encrypted or anonymized, it may lose its usefulness, rendering AI algorithms less effective or altogether unusable. Conversely, insufficient encryption or faulty anonymization can leave vulnerabilities that jeopardize personal privacy.

As AI integrates deeper into sectors like healthcare, finance, and personal services, the importance of protecting sensitive information escalates. Health data, for instance, contains some of the most private aspects of a person's life. AI's potential in revolutionizing healthcare is immense, but without robust encryption and effective anonymization, the intimate details of one's health could be laid bare to inscrutable elements of the digital world.

The development and enforcement of policies and regulations are also instrumental in guiding the use of encryption and anonymization in AI. Governance frameworks must enforce stringent standards for data protection to instill confidence that AI operates with the utmost respect for individual rights. This also means that industry practitioners must adhere to ethical considerations when implementing AI solutions.

Education plays a pivotal role in the adoption of encryption and anonymization methods. Professionals working in AI must possess apt knowledge and skills to securely manage data. This requirement spans beyond technical expertise to include an understanding of the ethical and societal implications of privacy breaches.

Moreover, public awareness about data privacy and the measures taken to ensure it is paramount. As users engage with AI-driven technologies, comprehending the safeguards in place, such as encryption and anonymization, empowers them to make informed decisions about their participation in such systems.

Looking at anonymization specifically, there's a continuous evolution of how anonymized datasets must be handled. Recent technological advancements enable de-anonymization through sophisticated algorithms and cross-referencing datasets, raising concerns about the long-term effectiveness of current anonymization techniques. Persistent innovation in this domain will be critical to ensure that anonymization remains a viable tool for privacy preservation.

As breaches and misuse of data continue to make headlines, trust in AI systems hangs in the balance. Encryption and anonymization are not just technical aspects of data management; they are foundational to the trust that society places in AI. Without them, the potential of AI could be significantly hamstrung by reluctance from the public and institutions to embrace these technologies.

In conclusion, encryption and anonymization are indispensable mechanisms within the ecosystem of AI. They are the silent guardians of privacy, working tirelessly in the background to enable AI systems to function while upholding the privacy rights of individuals. As we continue to innovate and fine-tune AI, the significance of encryption and anonymization will undoubtedly rise, remaining central to the discourse on the responsible and ethical use of artificial intelligence.

Chapter 6:
Human-AI Collaboration

In the wake of discussions on data protection and privacy, we now turn our attention to the synergistic potential of Human-AI collaboration. This chapter delves into the transformative ways in which AI can extend and enhance human capabilities, fostering an environment where the sum is greater than its parts. AI, when designed with ethical intent, becomes a power-multiplier for human potential, pushing the boundaries of productivity and creativity. Within these pages, we explore scenarios where AI systems work seamlessly alongside human intellect, bringing forward the dawn of augmented professions and enriched personal lives. The fusion of human intuition with AI's computational prowess opens new horizons for innovation, allowing us to solve complex problems with a level of precision and efficiency previously unattainable. Yet, as we chart this new territory, we also consider the long-term societal implications, striving to build a future where human collaboration with AI is grounded in mutual enhancement rather than displacement—where technology uplifts, empowers, and inclusively benefits all strata of society.

Enhancing Human Capabilities with AI stands as an invigorating chapter in the narrative of artificial intelligence, where technology's convergence with human skillset and intellect forms a symbiotic relationship, breeding a new era of potential. It is in this arena that we begin to witness the transformative power of AI to surpass the inherent limitations of our biology and psychology, creating a composite force with unparalleled capabilities.

The integration of AI in cognitive tasks illuminates this new frontier. Complex decision-making, once the sole preserve of highly-trained experts, can now be augmented with the analytical precision of AI, presenting us with opportunities to deliver solutions to problems with a speed and accuracy previously unattainable. In sectors such as finance, AI algorithms analyze vast arrays of data to predict market trends, empowering human financial advisors with insights that inform better, more strategic investment decisions.

In the medical field, AI's impact is no less profound. Surgeons operate with robotic precision, guided by intelligent systems that can reduce human error to near-zero levels. These AI-assisted procedures herald not just better outcomes for patients but also a shift in how medical practitioners are trained and how they refine their skills over time. AI does not replace the surgeon but rather enhances their capabilities, turning years of medical knowledge and experience into a collaborative effort with technology.

Moreover, the physical limitations that once seemed insurmountable are increasingly being transcended with AI's help. Exoskeletons and prosthetics integrated with smart systems are redefining mobility and dexterity. Individuals who lost limbs or were born with disabilities are discovering new independence, and the line where human capability ends and AI begins is becoming beautifully blurred.

Education, too, feels the touch of this collaboration. Personalized learning environments, powered by AI, adapt to the pace and learning style of each student, enabling a tailored education that optimizes learning outcomes. The AI acts as a personal tutor, identifying weaknesses and reinforcing concepts until mastery is achieved, thus enhancing the educator's role in developing potential.

AI also augments human capability in less visible, but equally significant ways. It is at work in language translations, allowing us to overcome language barriers and fostering global communication. It integrates with our devices, providing voice-controlled assistants and predictive text inputs, streamlining our interactions and bolstering our productivity.

The realm of creativity is not immune to the touch of AI. Artists and designers use AI tools to explore new aesthetic territories, creating artworks, music, and literature that merge human imagination with algorithmic complexity. The result is not only technically impressive but often evocative, pushing the boundaries of how we define creativity and its origins.

In the workplace, cognitive automation is transforming how we tackle routine tasks. Bureaucratic processes that consumed hours of human attention can now be executed in minutes by intelligent systems, enabling employees to focus on higher-level functions that demand human insight and creativity. This shift is not about replacing the workforce but about enriching jobs with more meaningful activities.

The progressive augmentation of our senses through AR (Augmented Reality) and VR (Virtual Reality) is yet another example of how AI enriches human experience. These technologies, backed by AI, allow for immersive experiences that enhance training, entertainment, and exploration of digital worlds that were once the stuff of science fiction.

AI-based safety systems in vehicles demonstrate how the technology can act as an extension of human reflexes and intuition, helping to reduce accidents and enhance driver capabilities. As autonomous driving technology evolves, it transforms the transportation landscape, melding machine precision with human oversight to create safer, more efficient roads.

Even in our homes, smart systems learn our preferences and adjust our environment to suit our comfort, allowing us to live more efficiently and with greater ease. They help manage energy consumption, provide security, and even anticipate our needs – an invisible yet invaluable augmentation to our daily living.

But with great power comes great responsibility. Ethical considerations must be at the forefront as AI continues to intertwine with human capabilities. The moral imperative to use such technology for the greater good and to ensure equity in its benefits is paramount.

Training and education must also evolve to keep pace with these changes. As AI redefines various roles, continuous learning becomes a necessity, ensuring that the workforce can transition smoothly into this AI-augmented world.

This pioneering era is not without its challenges, but the narrative of human progress has always been one of overcoming obstacles. The alliance between human intelligence and artificial intelligence promises to unlock new levels of achievement, and it's up to us to guide this partnership toward a future that reflects our highest values and aspirations.

The implications are vast and complex, yet the potential for good is enormous. As we unlock the capabilities of AI to enhance our own, we step into a future that is not just automated, but amplified – a world where our human potential is not replaced, but is instead unleashed in concert with the machines we've created. The future of this symbiosis is as exciting as it is uncertain, and it is our collective responsibility to help shape it for the betterment of all.

Ethical Design of Human-Centric AI As we delve into the vast and intricate domain of Artificial Intelligence, an essential topic that sits at the heart of responsible innovation is the ethical design of human-centric AI. The importance of creating AI systems that serve humanity's broad range of needs, and more importantly, align with our collective moral values, cannot be overstated. This emphasis is not only crucial for fostering trust and acceptance but also for ensuring the benefits of AI technologies are equally distributed across society.

The notion of human-centric AI involves a design approach that prioritizes human dignity, agency, and interests in the face of rapidly advancing AI capabilities. It is rooted in the understanding that technology should augment human potential, not undermine or replace it. As creators and regulators of AI, we must question not just what AI can do but what it should do to enhance the human experience.

When embarking on the design of AI systems, one must first consider the impact on human rights. The right to privacy, freedom of expression, and the right to be free from discrimination are foundational elements that must be embedded into AI systems. For AI to be truly human-centric, we must build these safeguards into every level of its architecture and operation. This layer of ethical foresight is key in advancing AI that protects, rather than jeopardizes, our shared human rights.

Transparency is another cornerstone of human-centric AI design. How can individuals trust decisions made by AI if they can't understand the processes leading to those decisions? This transparency extends beyond explaining the technical side of AI operations. It must include clarity in how AI will be utilized, the values it's programmed to prioritize, and how it interacts with other social and ethical standards in various contexts.

Accountability intersects with transparency and offers another layer of ethical consideration. Who is responsible when an AI system causes harm or acts in unexpected ways? A human-centric approach to AI ensures that accountability mechanisms are in place. This could mean establishing clear lines of responsibility within organizations that deploy AI, ensuring there's a human in the loop for critical decision-making processes, or developing robust auditing systems to monitor AI behaviors.

Incorporating inclusiveness during design processes is fundamental to create AI that serves everyone. Diverse teams should be involved in AI development to anticipate how technologies might impact different groups. Inclusiveness in AI design aids in detecting and minimizing bias that inadvertently arises from homogenous development teams overlooking the experiences of marginalized groups.

Bias and fairness must be considered throughout an AI system's lifecycle. From the initial data collection to the algorithms' coding and the final output, scrutiny is essential to avoid reinforcing societal inequalities. Ensuring datasets are diverse and representative while mitigating bias through algorithmic fairness techniques is crucial.

Empathy must also be infused into AI systems. While machines lack consciousness, human-centric AI necessitates consideration of human emotions and conditions. User experience should encompass emotional intelligence, such that AI systems recognize and appropriately respond to human emotions, enhancing the machine's role as a partner rather than merely a tool.

Respect for autonomy in AI systems centers on empowering individuals rather than controlling them. AI should support decision-making, provide options, and inform, but the final judgment must lie with humans, ensuring that AI remains a tool through which individuals exercise their own will and preferences.

Lastly, AI should strive to advance societal welfare. Its applications must be examined through the lens of the common good, aligning with goals that pursue environmental sustainability, social welfare, and economic prosperity. This broad view of AI's role encourages a holistic understanding of progress, where technology uplifts society as a whole and not just a select few.

However, ethical design isn't solely about achieving a set of static criteria; it's an ongoing process requiring constant vigilance and reevaluation. As societal values and expectations evolve, so too must the design and regulation of AI. Continuous dialogue among technologists, policymakers, ethicists, and the public is essential to recalibrate the moral compass guiding AI development.

Strengthening interdisciplinary partnerships is also necessary. Ethicists, sociologists, legal experts, and technologists must work together to balance the technical possibilities of AI with societal needs and ethical judgments. Such cooperation ensures a well-rounded approach to creating human-centric AI systems.

Moreover, education on ethical AI design must be incorporated into the curriculum for budding technologists and AI developers. Laying the foundation for ethical considerations early in their educational journey will help instill the principles necessary to create AI with human welfare in mind.

The policies and guidelines governing AI should reflect human-centric values. Legislation and industry standards must be drafted with the aim to protect individuals and communities, demanding that AI systems adhere to ethical norms before they are deployed. This enforcement is key to upholding high standards in AI development and use.

In conclusion, the fabric of human-centric AI is woven with threads of rights, transparency, accountability, inclusiveness, bias reduction, empathy, autonomy, societal welfare, adaptability, interdisciplinary collaboration, and principled governance. By adhering to these standards, we can envision and actualize a future where AI not only coexists with humanity but also strengthens, complements, and upholds the very essence of our human spirit.

Long-Term Societal Implications — the integration of artificial intelligence into the fabric of society casts a deep and enduring influence on how we live, work, and interact. The ripples of today's AI advancements will evolve into tidal waves of change in the decades to come. As we ponder the ramifications, it's not just about what AI will do for us, but also what it will do to us, our institutions, and future generations. To navigate these waters, we must consider the long-term societal implications with a sense of hope and a charter of responsibility.

The world we inhabit is standing at the cusp of an intelligence revolution. AI's ability to learn, reason, and adapt is not merely shifting power dynamics, but also redefining human identity and existence. The resulting societal transformation is inevitable and far-reaching. As with all epoch-defining technologies, we must tread the path of integration with careful consideration of long-term outcomes.

One such outcome, often debated and held under scrutiny, is the phenomenon of job displacement. Traditionally, automation and technology have tended to create more jobs than they replace. However, AI challenges this assumption with its capacity to automate cognitive tasks, forcing us to contemplate not only new employment paradigms but also the very nature of work. What sort of new occupations will emerge, and will they provide ample and meaningful opportunities for the evolving workforce?

The intertwining of AI within society also presents unique challenges to education systems worldwide. As AI technologies become more integrated into various industries, the skill sets required for tomorrow's workforce will change significantly. Education must evolve alongside these changes, fostering learning environments that emphasize creativity, problem-solving, and adaptability. Schools, universities, and vocational training programs will need to decipher and implement new methods to prepare students for an AI-rich future.

Similarly, the concept of 'value' in the context of societal contributions may be poised for a reexamination. AI's influence could lead us to value social connections, emotional intelligence, and ethical considerations more acutely. Moral questions, once the realm of philosophers, are becoming central to the governance of AI systems. It's essential to foster an environment where ethical considerations hold as much significance as technological advancements.

With AI, the potential for amplified inequality must be front of mind. While AI has the power to elevate our quality of life across numerous domains, it also possesses the capability to widen the chasm between the haves and have-nots. Addressing this inequality means ensuring that AI's benefits are distributed evenly across all strata of society, and not concentrated within the hands of a powerful few.

Moreover, we're witnessing the cultural metamorphosis influenced by AI-powered tools and platforms. The arts, languages, traditions, and even our perceptions are being shaped by the digital experiences created by AI. We may find ourselves at a crossroad where the essence of our cultural artifacts and shared experiences are co-authored by algorithms that learn from and respond to us.

Further consideration is warranted for AI's impact on governance and democratic institutions. With AI being used for everything from optimizing public services to predictive policing, the nature of public trust and the expectations of the governed will invariably shift. The use of AI in these contexts necessitates the utmost levels of transparency, accountability, and safeguards against misuse.

One cannot overlook the influence AI will have on human interactions and relationships. As AI becomes more sophisticated, it's conceivable that our interactions with it will become more personal, and our reliance on it more profound. This could change the way we perceive intimacy, trust, and companionship. How we prepare the upcoming generations for these shifts in social dynamics will be pivotal.

The long-term ecological implications of AI also demand attention. Beyond its potential to solve pressing environmental issues, the environmental footprint of AI itself—from massive data centers to the extraction of rare materials—must be managed sustainably. Our planet's health and the future of AI are interdependent. We must build systems that not only last but also foster the longevity of other life on Earth.

AI's rise raises complex geopolitical questions. With the 'global race for AI dominance', nations are incentivized to secure their interests through technology leadership. However, this competition, if left unchecked, could lead to an unstable international landscape. The consequences of a fragmented AI ecosystem, where global cooperation is truncated by nationalistic pursuits, are far-reaching.

There also lies the philosophical inquiry into the nature of consciousness and sentience, given the development of increasingly sophisticated AI systems. The way we address the rights and considerations for non-biological intelligences, should they ever reach a semblance of sentience, will reflect and redefine our values as a society. These debates will touch upon the most profound questions of what it means to be alive and to possess rights.

As much as we envision AI as a tool for advancement, it's also a mirror reflecting our biases, aspirations, and ethical contours. Its evolution will compel us to confront aspects of ourselves that may be uncomfortable or challenging. It's this reflection that will ultimately shape the norms and mores of future societies.

The long-term societal implications of AI don't solely rest on the shoulders of technologists and policymakers but involve every individual with a stake in the future. It's the collective responsibility of educators, artists, entrepreneurs, civic leaders, and citizens to steer this juggernaut of a technological phenomenon toward the best possible horizon.

The journey with AI is boundless and the destination uncertain. Yet, if history teaches us anything, it's that the human spirit coupled with ingenuity can harness the most disruptive of forces for the greater good. AI offers immense potential if we guide its trajectory with wisdom, compassion, and the foresight to protect the interests of all of society's members, present and future.

In conclusion, the long-term societal implications of AI are a tapestry of interwoven threads — economic, ethical, educational, cultural, ecological, geopolitical, and philosophical. Understanding and shaping these implications require foresight, interdisciplinary cooperation, and a commitment to the principles of equity and sustainability. AI is not just transforming the world; it's offering an invitation to reimagine and rebuild it. As we continue onwards, let's ensure that this ever-evolving partnership with AI serves to uplift humanity, preserve our planet, and enrich our collective future.

Chapter 7:
Artificial Intelligence as a Catalyst for Innovation

Building upon the foundational knowledge and multi-faceted impact of AI discussed thus far, Chapter 7 delves into artificial intelligence as an unprecedented force for innovation that propels industries into vibrant territories of discovery and creativity. AI is not just a tool but a partner in crafting solutions and designing the future, fueling a renaissance in various domains—from the arts to science, and from entrepreneurship to governance. It spurs a metamorphosis in product development, service enhancement, and problem-solving strategies, which are illuminated through compelling case studies. The collaboration between seasoned enterprises and agile AI startups heralds a paradigm shift in how intellectual property is both generated and protected in the digital age. These synergies not only redefine the competitive landscape but also emphasize the importance of unlocking AI's potential responsibly and imaginatively. As we traverse this chapter, the transformative power of AI becomes evident, reinforcing the notion that its role in fostering innovation is integral to propelling humanity towards a future brimming with promise and unprecedented potential.

Case Studies of AI-driven Innovations

In the realm of AI-driven innovations, there are numerous groundbreaking case studies that illuminate the power of artificial intelligence in transforming industries and even societies at large. One such innovation is the development of Deep Learning techniques used in machine vision for the rapid and accurate diagnosis of medical images, a task that would traditionally take medical professionals significantly longer to perform.

Another impressive innovation is the use of AI in optimizing logistics and supply chain management. Companies like UPS and Amazon have deployed AI algorithms to ensure efficiency in routing and delivery, greatly reducing fuel consumption and contributing to the reduction of environmental impacts. This kind of artificial intelligence implementation showcases how pragmatic applications of AI can have significant real-world benefits.

AI has also made significant strides in the realm of natural language processing (NLP). A project known as GPT-3 has demonstrated the ability to understand, generate, and translate human language with unprecedented sophistication. Language-based tasks, from real-time translation services to contextual customer support chats, are being revolutionized by technologies derived from this research.

Autonomous vehicles represent another domain where AI innovations have far-reaching implications. With automakers and tech giants like Tesla and Waymo at the helm, the pursuit of full autonomy in cars is pushing AI and machine learning to new frontiers. Their studies contribute important data about safety, urban planning, and the future of transportation.

In finance, AI-driven innovations have enabled advanced fraud detection systems through anomaly detection. For instance, Mastercard utilizes AI to analyze transaction data in real-time, flagging potentially fraudulent activity and reducing the incidence of credit card fraud significantly.

Moving from the engine of commerce to the fountain of creativity, AI is making waves in the art world as well. An AI named

AIVA (Artificial Intelligence Virtual Artist) composes original musical scores by learning from thousands of pieces of classical music, demonstrating AI's potential in creative industries usually thought to be reserved for human ingenuity.

The agriculture sector is similarly witnessing an AI revolution. By employing drones and AI-powered analytic platforms, farmers can now monitor crop health, optimize pesticide application, and predict yields more accurately than ever before. This advancement is not just enhancing efficiency but is also leading to more sustainable farming practices.

Speech recognition technology has also experienced a transformation, thanks to AI. Systems like Apple's Siri and Amazon's Alexa are becoming increasingly sophisticated, better understanding diverse accents, slang, and the context within conversations. These innovations enhance user experience and accessibility, changing how people interact with their devices.

AI has even entered the realm of law enforcement with predictive policing tools. These tools analyze historical crime data to forecast criminal activity, enabling law enforcement agencies to allocate resources more effectively and possibly prevent crimes before they happen.

Meanwhile, in the media and entertainment sector, companies are leveraging AI for personalized content recommendations. Netflix, for example, employs complex algorithms that analyze viewing habits to recommend shows and movies, thus increasing user engagement and satisfaction.

In the always vital healthcare sector, AI is being leveraged to predict patient risk and support clinical decision-making. One notable example is Google's DeepMind Health project, which analyzes medical records to predict patient deterioration, potentially saving lives through more timely interventions.

AI-driven innovations have also made inroads into the energy sector. For instance, Google's DeepMind has also developed AI that improves the energy efficiency of its data centers, leading to a significant reduction in energy usage and setting a precedent for other large-scale facilities to follow.

Customer service has been transformed by AI through the use of chatbots and virtual assistants. These AI-driven technologies provide rapid, around-the-clock customer support across a variety of industries, from e-commerce to banking, improving customer satisfaction and efficiency.

The manufacturing sector isn't left behind in the AI revolution. Robotics infused with AI algorithms have enabled more precise and efficient production lines. Automakers, for example, are integrating these smart robotics to improve the safety and customization of vehicles.

To sum up, the case studies of AI-driven innovations present a broad spectrum of achievements where AI is rapidly becoming an indispensable tool. These breakthroughs are not only a testament to the technological prowess of AI but also signal a turning point in its ever-growing integration into various aspects of our daily lives.

Collaboration between Enterprises and AI Startups is a vital and ever-evolving component of the innovation ecosystem that fuels technological advancement and economic growth. In an age where artificial intelligence is not just a buzzword but a cornerstone for competitive edge, enterprises across varied sectors are increasingly seeking partnerships with AI startups to reap the benefits of agility, innovation, and specialized expertise.

Such collaborations are fertile grounds for symbiotic relationships. Startups, often nimble and pioneering in scope, can develop cutting-edge technology rapidly, yet may lack the market access, capital, or robust operational infrastructure that established enterprises have built over decades. On the flip side, traditionally structured enterprises benefit from infusing their business models with the fresh ideas and swift adaptive technologies that startups bring to the table.

At the core, these partnerships typically focus on co-developing products, enhancing services, or tweaking business processes to be more intelligent and efficient. For instance, an enterprise in the retail sector might team up with an AI startup to implement advanced machine learning algorithms for personalized customer recommendations, inventory management, or supply chain optimization.

Such collaborations can unfold in various formats. One common approach is through corporate venture capital, where an enterprise invests in an AI startup in exchange for equity and, usually, a seat on the board. This investment often comes with strategic synergies, where the startup's technology is integrated into the enterprise's product line, driving innovation and offering a pathway to scale for the startup.

Another model is through strategic partnerships or joint ventures, where both parties maintain their individuality but come together to work on a shared project or goal. This could involve combining resources and expertise to tackle a specific market challenge or develop new products collaboratively.

Incubators and accelerators sponsored by large corporations offer yet another collaborative template. AI startups are given the resources, mentorship, and industry connections they need to develop their businesses rapidly. In return, the sponsoring enterprise often gains early access to revolutionary technologies and the opportunity to guide these advancements in directions that align with their strategic objectives.

It's important to note that for such partnerships to be successful, there must be a harmonious alignment of goals, culture, and expectations. While the enterprise may pursue stability, process orientation, and return on investment, the startup might prioritize innovation speed, technological breakthroughs, and market disruption.

A structured collaboration should therefore establish clear communication channels, transparent objectives, and flexible frameworks to accommodate the dynamism inherent to AI development. For example, agreements that allow for the iterative nature of AI product development, which often requires continuous tweaking based on real-world feedback, can be particularly effective.

Moreover, the data-driven nature of AI products might also raise concerns around data sharing, governance, and privacy. A successful collaboration will need to establish mutual trust and put in place robust data management protocols that satisfy both parties' regulatory and security standards.

Educational components can also play a significant role in these partnerships, with both startups and enterprises frequently engaging in knowledge exchange. Enterprises might offer industry expertise and business acumen, while startups can provide cutting-edge technological insights. This cross-pollination of knowledge can drive innovation and accelerate the development of AI technologies that are both groundbreaking and grounded in market realities.

But this is just the tip of the iceberg when considering the impact of collaborations between enterprises and AI startups. Along with technological advancements, these alliances are shaping the future of work, contributing to the emergence of new job roles and the re-skilling of existing workforces.

In some cases, these collaborations have even helped to democratize AI by making high-caliber AI tools and services accessible to a broader market segment. Smaller enterprises, and even individual consumers, can benefit from the trickle-down effect of these innovations, as advanced AI capabilities previously reserved for large corporations become more widespread and affordable.

One must not overlook the potential challenges and pitfalls associated with collaborations between enterprises and AI startups. Issues such as intellectual property rights, equity dilution, cultural mismatches, or misaligned objectives can derail promising partnerships if not adequately addressed from the outset.

Therefore, due diligence, a strategic alignment of core values, and a mutually beneficial agreement framework are essential commodities for these collaborations to thrive. As the AI landscape continuously evolves, so too must the nature of these partnerships, always adapting to new technological thresholds and market demands.

As we examine the canvas of corporate-startup collaboration, it is clear that these alliances are more than mere transactions or contractual engagements. They are dynamic, living bridges that connect the foundational strength of established businesses with the revolutionary potential of AI innovation. Through them, the promises of artificial intelligence are not just conceptualized but also actualized, heralding new horizons for businesses and society alike.

Intellectual Property and AI

As society delves deeper into the innovative realm of Artificial Intelligence, the intersection of intellectual property (IP) and AI presents a complex matrix of legal, technical, and ethical questions. The generation of new ideas, the creation of software, and even the artwork or music produced by AI systems raise important considerations about who - or what - can truly claim ownership. Intellectual property law, initially designed to protect human ingenuity, now confronts a novel challenge: the rise of the non-human inventor.

The significance of IP rights in AI is evident in the surge of AI-related patent applications. These are often submitted by corporations investing vast amounts in AI research, indicating the high stakes involved. Patent law, in this context, faces a fundamental question: can

AI-generated inventions be patented, and if so, who should be listed as the inventor? Traditionally conceived for human-created inventions, patent frameworks worldwide are grappling with this unprecedented situation.

Copyright questions are no less pressing. Creative outputs from AI, whether they be written work, visual art, or music, challenge the conventional notion of authorship. In most jurisdictions, copyright is reserved for human authors, yet when an AI generates artwork, the lines blur. What does this mean for creators who use AI as a tool, and how do such contributions compare to those crafted without AI's assistance?

Even trademark law, which protects brand identifiers from misuse, encounters novel issues in an AI-centric world. Machine learning algorithms might unintentionally infringe on existing trademarks by generating similar logos or brand names. It raises questions about liability and the extent to which AI creators can control or predict their creations' actions.

An intriguing development in AI's relationship with IP is the concept of 'AI as inventor'. Recent legal cases ask whether AI systems can be recognized as inventors on patent filings. This debate not only tests the legal definitions within IP law but also provokes reflection on the value we assign to human versus machine creativity and problem-solving.

Trade secret protection also intersects interestingly with AI. Companies may use trade secrets to safeguard their proprietary algorithms from competitors. However, the nature of AI and machine learning, which often requires datasets to be shared for further development, can sometimes conflict with the desire to maintain secrecy.

Licensing and the freedom to operate become complex when AI is involved in the innovation process. Organizations must navigate a maze of existing IP rights to ensure they have the freedom to use, develop, and commercialize AI technologies. Licensing agreements may require rethinking or new models to account for the uniqueness of AI systems.

The emergence of open-source AI also places a unique spin on IP considerations. Open-source AI projects, which make their code freely available for others to use and modify, complicate traditional IP paradigms focused on exclusivity and ownership. They thrive on community collaboration in lieu of proprietorship - a stark contrast to the competitive edge sought in patent protection.

Internationally, the legal landscape of IP rights in AI is highly fragmented. Different countries may take varied stances on AI and IP, causing potential confusion for multinational corporations and global initiatives. A universally agreed-upon framework seems distant, although international organizations like WIPO are initiating discussions to address these disparities.

Among all these concerns, there lies an overarching matter: the pace of technological advancement versus the speed of legal reform. The wheels of change within legislative bodies turn slowly compared to rapid AI developments. This dichotomy frequently leaves innovators in a state of uncertainty about how to protect their AI-driven technology adequately.

The notion of IP in AI also challenges existing business models. Companies must adjust their strategies to ensure they can both protect and monetize AI inventions. For instance, incorporating AI into products may require revisiting licensing agreements to spell out new IP considerations in distinctly uncharted territory.

For policymakers, AI's impact on IP law is a call to balance the promotion of innovation with the safeguarding of creators' rights. They must establish policy frameworks that recognize the fluidity of AI's creative potential while ensuring that human inventors are not disadvantaged nor disincentivized.

From a societal perspective, how IP is managed in the context of AI could significantly influence public access to information and technology. The ways in which laws and policies evolve could ultimately affect innovation and economic growth. It requires a keen understanding of both the technology's potential and the legal mechanisms at play.

In education and research, the implications for IP extend to how AI tools are used in the creation and dissemination of knowledge. Academia must navigate these waters carefully, fostering innovation while respecting the IP rights that come into play with AI-assisted research and development.

In conclusion, the nexus of intellectual property and AI is as fascinating as it is intricate. It demands a nuanced approach that is simultaneously rooted in legal precedent and forward-looking to adapt to a future where AI plays an ever-increasing role in creativity and invention. The next wave of innovation rests upon our ability to reconcile technological prowess with the imperatives of IP law, ensuring a future that respects both human and machine contributions to progress.

Chapter 8:
AI from a Global Perspective

As we delve into the intricacies of Artificial Intelligence through its astounding journey, it becomes crucial to widen our lens and capture the kaleidoscope of its impact on the world stage. This chapter guides us across continents, exploring how international policies and regulations serve as both enablers and barriers to AI's ascent. Here, we dissect the fascinating geopolitical dynamics that underpin the development of AI, as nations navigate between the fervent aspiration for technological advancement and the conscious efforts for maintaining digital sovereignty. We observe how the global race for AI dominance unfolds, with countries strategizing to harness AI's potential for economic growth, national security, and societal well-being while grappling with the complexities of competitive fairness and collaborative progress. Harnessing a global perspective allows us to appreciate the manifold realities of AI, as it emerges not merely as a transformative technological force but as a pivotal player in the intricate ballet of international relations and global power structures.

International Policies and Regulations In our interconnected world, artificial intelligence (AI) has transcended the boundaries of national jurisdictions, prompting a need for comprehensive international policies and regulations to manage its integration and ethical use. As we journey through the global landscape of AI governance, we encounter a patchwork of strategies, each shaped by unique cultural values, economic interests, and societal norms, revealing the complexity of achieving universal consensus on AI regulatory frameworks.

Why should we care about international policies and regulations for AI? Quite simply, it's about harnessing the transformative power of AI while safeguarding against the risks it poses to security, privacy, equity, and autonomy. As AI systems become embedded in everything from global financial systems to critical infrastructure, the stakes couldn't be higher. The governing regulations are an essential part of the fabric that will ensure AI benefits all of humanity without exacerbating global inequities or sparking uncontrolled adverse outcomes.

Navigating this terrain, it becomes clear that there is no one-size-fits-all approach to AI governance. The European Union, for instance, has adopted a value-driven approach focused on human rights and ethical principles. The EU's General Data Protection Regulation (GDPR) came into force in 2018, setting a global benchmark for data protection and privacy. It grants individuals significant control over their personal data and imposes strict rules on data processors and controllers, including those using AI.

In contrast, the United States, with its patchwork of federal and state laws, emphasizes market-driven innovation and has been slower to implement comprehensive AI-specific legislation. Instead, industry-led guidelines and standards often play a significant role, alongside specific regulations that address AI in the context of healthcare, transportation, or finance. The National Institute of Standards and Technology (NIST) is one of the bodies helping to shape policies that promote trust in AI through standards and guidelines that protect privacy and civil liberties.

China, a formidable player in the AI space, views AI development through a strategic lens, emphasizing state-sponsored initiatives and investment. The Chinese government released the "New Generation Artificial Intelligence Development Plan" in 2017, outlining an ambitious timeline to become the world leader in AI by 2030. Its approach to AI policy is characterized by national strategies that prioritize economic growth and technological prowess, often at the expense of individual privacy protections.

Moving away from specific countries, international collaborations are taking shape in the form of alliances and consensus-building efforts. The G7 nations, for example, have discussed guidelines for the ethical use of AI, although no binding policies have emerged. Meanwhile, UNESCO is working on a universal regulatory framework. They have drafted recommendations with the goal to provide a baseline for AI ethics that reflects international diversity and helps bridge regulatory gaps among nations.

Aside from government initiatives, multilateral institutions like the OECD have an important stake in the game. In 2019, the OECD adopted the Principles on Artificial Intelligence, the first international standards agreed upon by countries. These principles promote AI that is innovative and trustworthy and that respects human rights and democratic values.

The World Economic Forum is another international entity engaging various stakeholders to address governance gaps in AI. It launched the Centre for the Fourth Industrial Revolution Network, which brings together businesses, governments, civil society, and experts from around the world to co-design and pilot innovative approaches to policy and governance for new technologies.

Still, the fragmentation of international policies stands as a barrier to the seamless global operation of AI-based systems. Issues like cross-border data flows and the conflicting standards of privacy and data protection cause significant challenges for multinational corporations and SMEs alike. While some countries welcome the free movement of data as an economic catalyst, others see strict data sovereignty laws as essential for protecting their citizens' rights and security.

It's important not to overlook the role of trade agreements in shaping the AI landscape. Agreements like the United States-Mexico-Canada Agreement (USMCA) have set precedents by including digital trade provisions that affect AI, such as prohibiting data localization measures and protecting source code. These trade agreements sometimes become vessels for setting informal AI policies that can have widespread effects beyond the borders of the signatory countries.

As AI continues to evolve, one of the greatest challenges is keeping regulatory measures abreast of technological advancements. AI policies must balance the competing interests of stakeholders while relentlessly striving to remain adaptable and future-proof – no small feat given the rapid pace of innovation in the AI field.

Moreover, with the entrance of AI into high-stakes domains such as autonomous weapons systems, international security concerns come to the fore. Here, the United Nations plays a pivotal role through groups like the UN Group of Governmental Experts (GGE) on Lethal Autonomous Weapons Systems (LAWS). These platforms facilitate important discussions around the regulation of potentially disruptive technologies, though reaching consensus among member states has proven difficult.

Climate change is another area where international collaboration on AI regulation might have a profound impact. By setting global standards on how AI can be leveraged for environmental sustainability, nations could make significant strides towards the UN Sustainable Development Goals (SDGs). International AI applications in climate modeling and disaster prediction show the potential for AI to contribute to the welfare of the planet, but they also require cooperative international oversight.

In conclusion, as AI technology strides forward, the development of robust and comprehensive international policies and regulations becomes more critical. Such policies must be crafted in a participatory, multidisciplinary manner, considering a spectrum of voices and perspectives to ensure that the trajectory of AI aligns with our collective values and aspirations. Current efforts reflect only the beginning of an ongoing process that must adapt dynamically to the changing capabilities of AI, ever with an eye towards fostering innovation, ensuring security, and promoting universal ethical standards.

Without a doubt, AI has the power to reshape the world in myriad ways. But whether these changes lead to equitable and sustainable outcomes depends in large part on the international policies and regulations we put in place today. They serve not only as a compass for navigating the complex moral and ethical considerations AI poses but also as a blueprint for harnessing AI's transformative power to uplift and benefit societies across the globe.

Geopolitical Trends in AI Development The story of artificial intelligence is not just a tale of technological advancements and breakthroughs; it is deeply intertwined with the geopolitical fabric of our world. As nations recognize the strategic value of AI, it has become a significant element in the pursuit of economic, political, and military power. In this vital subsection, we will explore the complex landscape of geopolitical trends that shape and are shaped by the development of AI.

In the current climate, the United States and China are lead actors on the global stage of AI development. The United States, with its pioneering Silicon Valley culture, has long been at the forefront of innovation in technology. It has been home to major AI breakthroughs and houses tech giants that invest heavily in AI research. The federal government supports this endeavor with initiatives to ensure continued dominance in the AI field.

China, on the other hand, has made it a national priority to become the world leader in AI by 2030. To this end, the Chinese government has launched ambitious programs, providing substantial funding and policy support. China's approach to AI development is systematic and integrated, involving state-led initiatives that foster collaboration between researchers, industry players, and the military.

Beyond the US and China, the European Union is also positioning itself as a key player in AI, with a focus on ethical AI development. Emphasizing the importance of human-centric and trustworthy AI, the EU has introduced regulations and guidelines, such as the GDPR, that influence how AI is developed and used not only in Europe but globally.

Russia, while having fewer resources compared to the US and China, has signaled its intentions to be a formidable force in select areas of AI, particularly those related to defense and security. The country's leadership has openly recognized AI's strategic importance, especially in the context of military applications which could potentially shift the balance of power.

Emerging economies, such as India and Brazil, are not staying on the sidelines either. They are making strides in AI with the intention of leapfrogging to more advanced stages of economic development. By tapping into their growing pools of tech-savvy youth and entrepreneurs, these countries hope to carve out their niche in the global AI infrastructure.

One significant trend is the formation of international AI research alliances and partnerships. Such collaborations are becoming increasingly important, as they allow for pooling of resources, sharing of knowledge, and forging of strategic ties that bolster the capacities of member countries to innovate.

Another trend is the AI arms race, which has seen countries heightening their focus on autonomous weapon systems and surveillance capabilities. This militarization of AI raises numerous ethical concerns and prompts dialogues and potential treaties to regulate AI in warfare.

The AI development landscape is also marked by efforts of countries to secure their supply chains for critical components like semiconductors. As AI systems require advanced hardware, control over these supply chains becomes a geopolitical bargaining chip.

Intellectual property rights in AI are also turning into a contested geopolitical issue, with countries and corporations grappling over the ownership of AI-related patents and the setting of international standards in AI technologies.

Cybersecurity is another domain where geopolitical tensions manifest. The proliferation of AI has given rise to sophisticated cyber threats, leading countries to invest in AI-driven cybersecurity solutions and question the role of external AI systems in their critical infrastructure.

Digital sovereignty is gaining momentum, with countries seeking to control data flows and localizing data storage to protect their national interests. In this sphere, AI plays a dual role: as a tool for sovereignty and an asset that may require sovereign protection.

The scramble for AI talent fuels a global competition, leading to brain drain in some regions and the creation of AI hubs in others. Governments are now tasked with not just fostering AI education domestically but also crafting policies to attract and retain AI experts from around the world.

Environmental considerations are becoming a part of the AI geopolitical conversation as well. The immense energy demands of training complex AI models have prompted nations to look for sustainable ways of advancing AI without compromising environmental goals.

Lastly, international diplomatic relations are increasingly colored by AI. Countries use AI as an instrument of soft power, providing aid in AI initiatives, training foreign AI experts, and facilitating the adoption of AI technologies abroad to strengthen political ties and expand influence.

The exploration of these geopolitical trends in AI development reveals a multi-faceted web of strategic, economic, and social factors that are inextricably linked. Nations must navigate this terrain with foresight and responsibility, as the decisions made today will shape the global dynamics of tomorrow. AI is not just a transformative technology; it is a central axis around which the future of global power will revolve.

The Global Race for AI Dominance As we delve into the complexities and triumphs of artificial intelligence, it becomes increasingly clear that AI isn't just about technological terabytes and circuitry. It's also about global leadership and economic prowess. The quest for AI supremacy has turned into a marathon where nations sprint. It's a race defined by both the collaboration of bright minds and the competition between economic powerhouses.

The seeds of this race were sown as governments recognized the transformative potential of AI. It was evident that whoever leads in AI will influence the future of industry, military power, and the societal structures at large. Understandably, countries are now fiercely investing in AI research and development, aiming to implement AI in areas ranging from healthcare to national defense.

In this teeming competition, the United States has made notable strides with Silicon Valley's tech giants pioneering groundbreaking AI advancements. These advancements are not just in digital personal assistants or search algorithms. They're in complex fields such as quantum computing, autonomous vehicles, and advanced machine learning frameworks too.

Nevertheless, global competition isn't far behind. China, with its ambitious AI development plan, aims to become the world leader by 2030. This plan isn't just about advances in the technical field; it's a concerted effort involving policy support, funding, education reform, and international collaboration.

The European Union, meanwhile, takes a strategic approach that emphasizes ethical AI. It seeks to compete on the global stage by establishing robust regulations and frameworks that ensure AI systems developed and used within Europe are trustworthy and respect privacy and human rights.

Russia too recognizes the strategic importance of AI. With national pride and security at stake, its approach leans heavily towards developing AI in defense and military sectors. As part of this, substantial support is directed towards both governmental and private sectors engaged in AI.

India, with its vast population and growing technical expertise, cannot be underestimated in this race. While perhaps a step behind China and the US in terms of investments and infrastructure, India is rapidly making inroads through initiatives aimed at fostering AI innovation and application in various sectors.

Countries like South Korea, Japan, and Canada are also making significant investments, focusing on aspects of AI that complement their existing economic strengths. South Korea in robotics, Canada in AI research and ethical AI, and Japan in AI application across industrial automation.

Such a race is not without its hurdles, though. Issues of AI arms races, particularly in the military domain, raise concerns about a new type of cold war - one possibly marked by cyber capabilities and autonomous weapons systems. It's a daunting thought that places an even greater emphasis on international cooperation to set boundaries and norms.

This competitive atmosphere has spurred an unprecedented level of collaboration between academia, industry, and governments. Through partnerships, we see an acceleration in the development of AI technologies far beyond initial individual capabilities.

The drive for dominance also fosters rapid innovation. Countries are vying to host the next big AI startup or to develop the next revolutionary algorithm. In doing so, they are creating ecosystems that support creative thinking and the rapid translation of research into real-world applications.

However, not all is about racing ahead. There's also a significant push for inclusive and responsible AI. The concept of leaving no one behind is promoting initiatives that ensure developing nations aren't left out of the benefits AI can offer. Global organizations and coalitions are working to democratize AI, ensuring it's a force for good and accessible for all.

At the heart of this race, talent is the most sought-after resource. Countries are revising their education systems, creating new visa categories for skilled AI professionals, and investing in retraining their workforce to nurture a pool of talent that can drive AI innovation.

The outcomes of this global competition will shape the future. They'll determine not only which countries lead the world in technology and economy but also how AI is integrated into our everyday lives. It's a race that goes far beyond mere technology – it's about vision, strategy, and the shape of the world to come.

As this race unfolds, we mustn't lose sight of the broader goal: harnessing the power of AI to solve some of humanity's most pressing challenges while ensuring that we maintain ethical standards and equality. The race for AI dominance is indeed a testament to human innovation and determination, illustrating that when it comes to the future, the sky isn't the limit – it's the starting line.

Chapter 9:
The Effects of AI on Daily Life

Emerging from a global landscape peppered with policies and regulations, we delve into the intimate nooks of our everyday existence to unravel the tapestry of change woven by artificial intelligence. AI isn't just an external force; it has seeped into the rhythm of our daily routines, reshaping the way we interact, learn, and tend to our health. It's in the personalized recommendations that greet us each morning, the intelligent assistants that orchestrate our schedules, and the seamless integration within our vehicles that guide us safely to our destinations. Healthcare has been revolutionized with predictive analytics for better diagnosis and treatment plans, while children and adults alike experience the transformative educational tools that adapt to individual learning paces. As we go about our lives, AI subtly reinforces our decisions, offering convenience but also challenging us to reconsider the evolving dynamics between humans and machines. This chapter delves into these intricacies, offering a crisp reflection of a world where AI is not a distant concept but an entwined partner in the dance of daily life.

Personalized Experiences through AI are not simply a matter of preference; they represent the evolution of technology's role in serving users in the most tailored and intuitive ways imaginable. Through artificial intelligence, we're entering an era where services aren't just provided; they're crafted in real-time to fit the nuances of individual needs and desires. We've seen the groundwork laid by personalization in online shopping recommendations and content streaming services. Still, AI's potential to offer personalized experiences stretches far beyond these early applications.

Consider the world of education, where AI can analyze a student's learning habits, strengths, and areas for improvement to provide a tailored educational pathway. Imagine a curriculum that adapts in real-time to maximize a student's potential. This isn't just a dream; it's swiftly becoming reality as classrooms around the globe begin to employ AI tools to cater to each individual's learning journey. What's remarkable is the capacity AI has to not simply assist but to enhance the educational process, leading to outcomes previously unattainable through traditional methods.

Personalization also radically transforms healthcare. AI systems can analyze vast amounts of medical data to offer individualized treatment plans, take into account a patient's genetic composition, lifestyle, and other variables. Such deep personalization can significantly increase the effectiveness of treatments and even anticipate health issues before they arise, fostering a proactive rather than reactive approach to healthcare.

In the realm of entertainment and media consumption, we're already witnessing the first waves of what personalization means. From music services learning our listening preferences to create the perfect playlist, to smart TVs that know which shows we'll binge next, AI-driven personalization is becoming the norm. This shift not only increases user satisfaction but also opens up novel avenues for content creators to match their offerings with the precise audience that will cherish them the most.

Personalization through AI even extends to customer service, with chatbots and virtual assistants now capable of understanding context and history with a customer. This allows these AI-powered solutions to offer a level of service that feels attentive and unique to each interaction, scaling customer service to new heights of efficiency without compromising the quality of care.

Let's also consider the personalization of products. AI is making it possible for companies to tailor products to individual specifications at scale. This could mean custom-fitting clothing manufactured on-demand or nutrition supplements blended daily to suit one's specific dietary needs. The crux of AI's impact here is its ability to make mass personalization economically viable where it was once only feasible for luxury goods.

In terms of home automation, AI is refining our living environments to suit our preferences in ways we may not even consciously notice. From lighting that adjusts to our moods to thermostats that learn our schedules and preferences for comfort, AI in smart homes makes our personal spaces even more intimately ours. AI doesn't just make life easier; it makes our homes an extension of our habits and preferences.

Turning our attention to marketing and advertising, the influence of AI is shifting the landscape of how consumers relate to brands. Personalized marketing isn't just about showing the right ad at the right time; it's about crafting a narrative and an experience that resonates on an individual level. AI's ability to interpret large datasets allows for pinpoint accuracy in delivering personalized messaging that aligns with the consumer's journey and intent.

The financial sector hasn't been left untouched by the wave of personalization either. AI-driven financial advice is customizing the way individuals and businesses plan and manage their finances. With personalized insights and recommendations, AI is enhancing financial literacy and empowerment—a key to making sound financial decisions in an increasingly complex world.

Travel and hospitality are also benefiting from AI's touch of personalization. From booking platforms that suggest destinations based on past trips to hotel rooms that adjust to a guest's preferences upon their arrival, AI is paving the way for experiences that feel uniquely designed for each individual traveler.

But all these personalized experiences come with a vital need for responsible data handling. As we delve into deeper layers of personalization, the sensitivity of data being analyzed and utilized by AI systems heightens. Ensuring privacy and security is paramount, as AI works best when it has the most accurate data—and that often means the most personal data. Balancing the personalization benefits with data protection is an ongoing challenge that requires diligent attention and innovative solutions.

Moreover, the advent of AI personalization invites us to reflect on the broader implications of such technology. How does the constant tailoring of experiences affect our exposure to diverse perspectives and challenge our preconceptions? An AI designed to accommodate our every preference might inadvertently lead to a narrowing of our worldviews if not managed with care.

In the end, AI's role in providing personalized experiences is transformative. It holds the power to touch the core of our daily lives, making experiences more enjoyable, educational encounters more enriching, and our overall lives more aligned with our individual needs and aspirations. As we continue to unlock AI's potential, we're tasked with the responsibility of guiding its ethical deployment for the betterment of all.

As such, the success of personalized experiences through AI is contingent upon the synergy between innovation, ethical AI practices, and regulatory frameworks that protect individual rights without stifling progress. Striking this balance is not only crucial but attainable, and it will define the quality of our personalized future.

Conclusively, with the tools and knowledge at our disposal, the potential to craft a world where AI supports everyone in a profoundly personalized manner is not only within sight but an achievable reality. It compels us to push the boundaries of AI's abilities while maintaining an unwavering commitment to upholding ethical standards, data privacy, and the inclusion of diverse perspectives.

Through these efforts, AI's promise of personalized experiences has the potential to become one of the most empowering technological advancements of our time.

AI in Health, Education, and Transportation

In the realm of health, artificial intelligence has become both a precision instrument and a formidable force in disease prediction, prevention, and management. Leveraging machine learning algorithms, AI systems are increasingly adept at parsing vast datasets of medical records to uncover patterns that human clinicians might overlook. Decision support systems, for instance, assist in diagnosing illnesses based on symptoms, medical imaging, and genetic information, demonstrating an accuracy that augments the expertise of medical professionals.

Moreover, AI-powered applications are revolutionizing personalized medicine, enabling treatments that are specifically tailored to an individual's genetic makeup. This advancement not only enhances the efficacy of interventions but also mitigates adverse drug reactions. Additionally, AI is instrumental in drug discovery, shortening the development cycle of new medications by predicting how different drugs will interact with targets in the body.

In the educational sphere, artificial intelligence is bridging gaps between diverse learning needs and the one-size-fits-all approach that has traditionally defined the sector. Adaptive learning technologies are forging paths towards personalized education, where AI algorithms adjust learning materials to match the pace and style of each student. This caters to individual strengths and weaknesses, potentially leveling the playing field for those who might struggle in a conventional classroom setting.

Robotics in education is not just a tool for engagement; it teaches coding and complex problem-solving skills critical in a technology-driven world. AI also alleviates administrative burdens, automating tasks like grading and record keeping, which grants teachers more time to focus on direct student interaction and instruction.

When it comes to transportation, artificial intelligence is a key driver of change, propelling the industry towards greater safety and efficiency. Autonomous vehicles, employing AI, sensors, and real-time data, are poised to transform how we commute by reducing human error—the leading cause of traffic accidents. Furthermore, AI applications in traffic management analyze patterns to optimize signal timings and reduce congestion, paving the way for smoother and faster travel within urban settings.

AI has stepped into another critical role within transportation—predictive maintenance. By analyzing data from various sensors on vehicles, AI can predict equipment failures before they happen, ensuring the reliability of everything from personal cars to large fleets of commercial vehicles. This not only saves costs but also greatly enhances safety on the roads.

The integration of AI also extends to air travel, where predictive analytics improve operational efficiency by forecasting maintenance issues and optimizing fuel consumption. Such applications lead to reduced delays, increased safety, and a decrease in the carbon footprint of the aviation industry.

In public transportation, AI enables smart routing and dynamic scheduling, matching supply with fluctuating demand and thus improving the commuter experience. Beyond just convenience, these smart systems potentially make public transportation more sustainable and enticing, a crucial step in reducing carbon emissions and tackling urban sprawl.

AI's influence in healthcare continues with robotic-assisted surgeries that combine the dexterity and precision of machines with the judgment and expertise of human surgeons. These systems enhance the surgeon's capabilities, potentially leading to less invasive procedures, reduced recovery times, and improved surgical outcomes.

Telemedicine, emboldened by AI, is another growing domain that offers diagnoses and treatment plans from afar, making healthcare accessible even in remote areas.

Educational platforms driven by AI not only improve individual learning experiences but also provide valuable insights for educators by analyzing student performance data. By identifying trends and potential issues, such as a common misunderstanding of a concept, teachers can adjust curriculum or offer targeted support. This data-driven approach ensures that educational strategies are responsive and effective.

The marriage of AI and health leads to systems that can monitor patient health in real-time, offering alerts for irregular patterns that may indicate a need for immediate medical attention. Wearable technology, equipped with AI, thus becomes an ongoing health assessment tool, fostering preventative care that can save lives and healthcare costs.

In education, virtual assistants powered by AI can handle routine inquiries from students, providing instant responses and freeing up human resources for more complex support tasks. This not only streamlines the administration but also tailors the educational experience to facilitate seamless student support services outside the traditional classroom hours.

AI's impact on transportation resonates through logistics and supply chain management as well. It predicts delivery times, optimizes routes, and manages inventory through real-time data analysis, enabling businesses to operate more efficiently and respond rapidly to market demands. This kind of predictive power fosters a resilient supply chain capable of adapting to disruptions swiftly.

The transformation brought by AI across these three sectors— health, education, and transportation—illuminates just a fraction of its potential. In healthcare, AI stands as the sentinel for our well-being; in education, it acts as a personalized mentor; and in transportation, it is the unseen navigator ensuring our journey is safe and smooth. The intersection of these arenas with AI not only paints a picture of an optimized world but also poses profound questions about the symbiosis of human intelligence with its artificial counterpart.

The synchronization of AI within these critical components of society blazes a trail for a future that is not only cognizant of the individual needs but is also structured to act upon them with unprecedented precision and care. As we continue to unwrap the layers of AI's capabilities, we inch closer to a world that harnesses this powerful technology to foster health, knowledge, and connectivity.

Concurrently, the advances in AI in these domains demand a vigilant approach to oversight, ethical considerations, and ongoing adaptation of regulations. Integrating AI isn't just about harnessing technological innovation—it's about shaping a society that is equipped to make the most of it, ethically, equitably, and sustainably. Such integration aligns with the goal of this book to herald not just the marvels that AI offers but the responsibilities it entrusts upon us. Thus, the convergence of AI in health, education, and transportation stands not only as a testament to human ingenuity but also as a beacon for continued collaborative evolution between humans and machines.

The Evolving Human-Machine Relationship has emerged as one of the most profound developments in the tapestry of our society, shaped by the relentless evolution of artificial intelligence. As we navigate through the complexities of this new era, the intertwining of human cognition with the precision of machine intelligence has begun to redefine what it means to work, learn, and interact in the interconnected world we inhabit. This chapter delves into the metamorphosis of our association with technology, an exploration that invites us to reflect on how we adapt and harmonize with the digital counterparts that are increasingly embedded in the fabric of our daily lives.

The relationship between humans and machines had its genesis in simple tools designed to augment physical abilities, but as AI continues to surge forward, we are witnessing a transformative collaboration that extends beyond the mere enhancement of physical tasks. AI systems today are not only executing complex calculations and processes but are also capable of learning from interactions, encompassing aspects of our intellectual and even emotional experiences. As such, our bond with these intelligent systems is acquiring unprecedented depth and intimacy.

To comprehend the current state of human-machine relations, one must appreciate the nuances of human-AI collaboration, which is characterized by a symbiotic dynamic. The advent of cognitive technologies has paved the way for machines capable of understanding our needs and behaviors, thereby seamlessly integrating into our personal and professional lives. These technologies assist in decision-making, provide personalized recommendations, and help us navigate through vast amounts of data, highlighting an evolution from assistance to partnership.

Our workplaces have transformed into hubs of human-AI synergy, where algorithms and automation complement human skills, leading to greater efficiency and creativity. In industries like healthcare, AI-driven analytics assist doctors in diagnosing diseases with greater accuracy, exemplifying a paradigm where machines and professionals coalesce their expertise for superior outcomes.

However, the burgeoning intertwine of AI extends far beyond the realm of employment. Education systems are deploying AI to provide customized learning experiences that adapt to the pace and style of individual learners, embodying a level of personalization that was previously unattainable. Students now find themselves interacting with intelligent tutors that cater to their unique educational needs.

At home, smart devices have graduated from performing basic tasks to understanding user habits and preferences, creating an ecosystem that anticipates and responds to the occupants' lifestyles. This technology not only simplifies routine tasks but also supports individuals in managing their health and wellbeing, showcasing the potential of AI to enhance the quality of life.

Moreover, the progression of AI has ignited a conversation about forming emotional connections with machines. With the emergence of social robots and virtual assistants that simulate conversation and companionship, we are venturing into uncharted territories of attachment and interaction. This development invites us to revisit our definitions of companionship and support, reflecting upon the psychological dimensions of our relationship with machines.

In the consumer landscape, AI has revolutionized the way we engage with brands and services. Personalization engines drive unique user experiences by analyzing behavior and preferences, providing a touch of individuality to the commercial exchange. This recalibration of client-service dynamics illustrates the way AI is personalizing the marketplace, aligning products and services more closely with consumer desires.

As advanced as these relationships are today, the trajectory suggests that they will become more complex and integrated. Anticipatory AI systems are already on the horizon, poised to detect our needs before we explicitly express them, further blurring the lines between proactive human requests and passive machine predictions. It's a future where AI doesn't just respond; it anticipates.

This integration raises critical questions about dependency and the delegation of responsibility to AI. How much control are we willing to transfer to these intelligent systems? Does this mark the dawn of a new era of passivity, or does it free us to pursue more creative and fulfilling endeavors? As we grapple with these decisions, our societal values and principles will invariably shape the boundaries of this relationship.

Equally important is the dialogue concerning the ethical design of AI systems. With an increasing degree of interaction and reliance on these technologies, ensuring that they reflect equitable and inclusive values is crucial. The task at hand is to instill ethical principles in these intelligent systems that safeguard against bias, which could otherwise perpetuate societal inequalities.

Conversely, as machines become more like us, there's an exploration of whether humans should become more like machines. The burgeoning field of human augmentation and biohacking is evidence of our quest to enhance our own capabilities. Should we see these efforts as an extension of the evolving human-machine relationship, or as a distinct trend in the pursuit of individual enhancement?

As we peer into the future, we can foresee that AI will play a defining role in shaping our societies. Its integration into governance systems, for instance, could lead to more informed policymaking and public administration. In public safety, AI can offer predictive insights that preempt threats and crises, potentially transforming the way we prepare for and respond to emergencies.

Yet, for all its potential, the narrative of the human-machine relationship is inextricably linked with the notion of trust. Our willingness to rely on AI is contingent upon the transparency and accountability of these systems. There is a looming imperative for technology creators to ensure that AI is not only robust and reliable but also understandable and explainable to the general populace.

In conclusion, tracing the trajectory of the human-machine relationship unfolds a history of mutual enhancement and an intricate dance of balance. It is a narrative of ingenious technology woven into the tapestry of human experience, a venture understood not solely through the lens of technological achievement but as an evolutionary leap in our social fabric. As we advance, it remains pivotal to foster a world where AI serves as an ally to humanity, propelling us towards a future marked by shared prosperity and collective wisdom.

Chapter 10:
Preparing for the AI Future

As we pivot from exploring the profound effects of AI on our daily lives, we must now turn a critical eye towards the horizon, contemplating the necessary steps to thrive in an AI-infused future. No stone should be left unturned as we scrutinize the educational paradigms that must evolve, inspiring ardor in the hearts and minds of future generations to embrace AI not as a distant concept but as a tangible thread in the fabric of their vocations. The onus falls on both governments and individuals to develop robust strategies that ensure readiness—not just as a pursuit of knowledge, but as a comprehensive approach that encompasses economic, social, and ethical facets of existence. The roadmap to such preparedness must be forged with foresight, inclusivity, and adaptability at its core, ensuring all levels of society can navigate the shifting tides of change. Above all, the preparedness for an AI future challenges us to reimagine our roles in a world where human intelligence and creativity align with the computational acumen of machines, promising a symphony of human-AI collaboration that amplifies our potential to reach unprecedented heights.

Educational Reforms for the AI Era

Educational Reforms for the AI Era are essential to ensure society can thrive alongside ever-advancing technological counterparts. These reforms are not just a suggestion; they're an imperative. The AI era has swept in with tides of change that have already begun to reshape the landscape of work, ethics, and daily life. Consequently, education systems worldwide must undergo a transformation to not only coexist with Artificial Intelligence but to harness its potential, nurturing a generation that is as adept and agile as the technology they will interact with.

First and foremost, there must be a recalibration of the curriculum. Traditional education has emphasized memory and routine skills, but AI's strength lies precisely in these areas. Hence, the new focus should be on developing cognitive flexibility, problem-solving skills, and emotional intelligence—a suite of abilities that AI can't easily replicate. Encouraging creativity and critical thinking from an early age will allow future generations to excel in tasks where human ingenuity is required.

Additionally, literacy in the digital realm must become a cornerstone of education. Understanding the principles of coding and algorithmic thinking isn't just for computer scientists but for all students. This digital fluency will empower students to be artisans of the AI world, capable of shaping technology to serve human needs rather than being passive consumers of it.

Interdisciplinary study is also crucial in the AI era. The integration of science, technology, engineering, and mathematics (STEM) with the arts and humanities acknowledges that the challenges solved by AI are not just technical but also ethical and societal. This holistic approach to education fosters a more nuanced perspective on the implications of AI advancements.

Collaborative learning environments must replace the competitive, individualistic classrooms of old. AI excels in individualized tasks but teaching teamwork and collaboration prepares students to work alongside both human and machine partners. These skills are vital in a future where AI tools and team members will likely be part of the workforce.

As AI personalizes learning experiences, the role of educators transforms. Teachers will become more like facilitators or mentors, guiding students through personalized learning paths carved out with the help of AI, rather than being the sole source of knowledge. This requires significant professional development and a shift in teaching methodologies.

With the ever-present threat of obsolescence, lifelong learning becomes a necessity rather than an option. Educational reforms for the AI era must advocate for learning models that support continual education beyond formal schooling, in recognition that learning is an unending journey in an ever-changing world.

Assessment methods, too, must be revolutionized. The focus should move from rote memorization towards project-based and experiential learning assessments. Showcasing one's ability to apply knowledge in practical, real-world situations is far more indicative of future success than traditional exams.

The impact of AI on education is not just methodological but also infrastructural. Classrooms should be tech-friendly spaces where the latest AI tools are both abundant and accessible. Equipping schools with modern technology is an investment in the future, enabling hands-on experience with tools that students will encounter in their professional lives.

Inclusivity must also be addressed in educational reform. Closing the digital divide by ensuring all students have access to quality AI education is crucial to prevent further societal stratification. This not only means access to technology but also to high-quality instruction, regardless of geographic or socioeconomic status.

Primary and secondary education reforms set the stage, but higher education must also adapt. Universities should focus on multidisciplinary AI research and development while aligning degree programs to the new market demands. This could mean the birth of new fields of study and the redefinition of existing ones.

Next, ethical training in AI must be pervasive across all levels of education. As much as students are taught how to utilize AI, they should also be instilled with a strong sense of ethics to ensure that they are guiding AI applications in ways that are fair, unbiased, and for the betterment of humanity.

The concept of 'global classrooms' should also be embraced, where students from different parts of the world learn with and from each other using AI and digital media. This prepares them for a workforce that is not only interprofessional but also international.

Further, education policy-makers and stakeholders must actively engage in discussions about the future of AI, ensuring that reforms are not reactive, but proactive. They must anticipate the ways in which AI will continue to evolve and pre-emptively create educational environments that can adapt to those changes smoothly and swiftly.

Finally, these educational reforms require collaboration between governments, educational institutions, the tech industry, and communities. It's a convergence of interests and expertise that will forge a robust educational framework, capable of sustaining the AI era and maximizing its potential for all.

In conclusion, as AI intertwines with the fabric of society, the reform of educational systems around the world is not just beneficial but absolutely essential. The generation we are educating today will be the pioneers of an AI-enriched world, and it's our responsibility to equip them with the knowledge, skills, and ethical grounding to navigate it successfully and humanely.

Anticipating the Needs of Future Generations takes foresight and a profound commitment to ensuring that our technological advancements pave the way for a thriving, equitable world. This goal is particularly significant as we delve into the realm of artificial intelligence—a transformative force poised to redefine every aspect of our existence. The children of tomorrow will live in a world deeply intertwined with AI. So, how can we mold this technology to foster an environment that encourages growth, equality, and sustainability?

In fashioning AI systems, it's integral to consider long-term consequences alongside immediate benefits. Short-sighted innovation might offer temporary advancements, but could also lead to unforeseen challenges that burden future generations. Wholehearted efforts must be directed towards building AI that aligns with the enduring values of society, with an emphasis on long-term wellness over ephemeral gains.

The AI we develop today will significantly impact education systems. We must weave AI into curriculums not just as a subject of study, but as a personalized tool to enhance learning. AI can cater to individual learning styles, pace, and preferences, potentially reducing educational disparities and lifting entire populations to new levels of literacy and critical thinking.

Reflecting on environmental sustainability, AI has the potential to monitor and manage natural resources with unprecedented precision. However, it's vital to ensure these intelligent solutions do not come at an exorbitant cost to the planet. Therefore, constructing eco-friendly AI—systems that are energy-efficient and minimize environmental footprint—is an obligation to our successors.

It's crucial to instill values of fairness, diversity, and inclusion within AI's very coding. Today's algorithms are tomorrow's judges, advisors, and companions. They must reflect the rich tapestry of human culture, sensitive to the nuances and needs of all sections of the population. This is not just a technical challenge, but one deeply-rooted in social values as well.

There is a growing need to consider how AI will alter the landscape of work. While automation will likely phase out certain jobs, fostering a culture of lifelong learning can empower individuals to pivot into roles that AI will create. The education systems of today must adapt to prepare students for an agile, AI-augmented workforce.

The ethical dilemmas that AI presents—like privacy concerns, surveillance, and decision-making transparency—demand that we provide robust frameworks for our children to navigate. We should build toward legislative bodies and watchdog organizations that can effectively oversee AI, ensuring it operates within the bounds of established social contracts.

Security in an AI-driven future is imperative. We're not merely protecting data and systems from cyber threats, but also safeguarding the very essence of human autonomy. Advancements in AI should be coupled with advancements in cyber defense strategies, protective technology, and public awareness about digital security hygiene.

As we make strides in AI-driven healthcare, it's essential to prioritize accessibility to these life-altering technologies. AI could revolutionize diagnosis, treatment, and preventive care, reducing the bureaucratic load on healthcare systems and leading to more equitable health outcomes. Yet, this must be strategically planned to avoid creating another facet of the digital divide.

Transportation is yet another domain where AI can yield benefits for future populations. Autonomous vehicles promise to revolutionize the ways in which we travel, reducing accidents and optimizing traffic flows. The infrastructure and regulatory frameworks for these technologies need to be laid now, with a keen eye on safety, efficiency, and urban planning.

To ensure responsible AI development, promoting public awareness and engagement is just as important as technical innovation. The citizens of tomorrow must be equipped with a sound understanding of AI, as well as the tools to question and shape its trajectory. This includes fostering critical thinking and ethical questioning from a young age.

Addressing the digital divide is a pressing issue, as AI has the power to either widen or bridge this gap. Inclusive access to technology can help flatten inequalities, providing future generations with a level playing field. It's our duty to establish infrastructure and programs that support equal access to AI tools and education.

Furthermore, AI's effect on our psychological and emotional well-being can't be overlooked. As AI becomes more embedded in daily life, it's crucial to guide future generations in managing their relationship with technology, endorsing balanced usage that promotes mental and physical health.

Anticipating the art-cultural implications of AI is a unique challenge. AI is not only a scientific endeavor but also a catalyst for new forms of creative expression. We must foster an environment that encourages the interplay between human creativity and AI, preserving cultural heritage while embracing innovative art forms.

Last but not least, envisioning AI's role in governance is key. Will it be a tool of empowerment, providing citizens with a voice and enhancing democratic processes? Or could it become a means of control? Crafting policy today that aligns AI with the principles of democracy will safeguard the interests of future citizens, allowing them to inherit not just technology but the value system that guides it.

AI's story is still unfolding, and we have a unique opportunity to shape its narrative. What we set in motion now will resonate through the ages. Let us dream of a future graced with AI's potential and take on the mantle of responsibility to build toward a world where technology not only advances our capabilities but also upholds our shared humanity for all who come after us.

Readiness Strategies for Governments and Individuals As we delve deeper into the Artificial Intelligence era, preparedness is paramount. Both governments and individuals must develop strategies to adapt to and reap the benefits of AI's transformative power while mitigating its risks and challenges. These strategies aren't just an option; they're a necessity in harvesting the positive fruits of AI while minimizing potential disruption.

For governments, readiness begins with public policy and investment. There needs to be a foundational layer of AI-friendly policies that encourage innovation while safeguarding the public. Governments should invest in infrastructure, such as high-speed internet and data centers that can process and analyze large amounts of data. This creates a fertile ground for AI technologies to grow and evolve effectively within their jurisdictions.

Education systems must be reevaluated to ensure they reflect the skills necessary in an AI-driven economy. Critical thinking, problem-solving and adaptability should be the focus, alongside technical skills in data science and coding. Forward-thinking curricula, apprenticeship programs and continuous education opportunities can help the workforce stay relevant in an ever-changing landscape.

Additionally, social welfare policies will need to evolve. With the potential for job displacement due to automation, strategies such as universal basic income or safety nets for retraining should be discussed and tested. Governments must also encourage entrepreneurial initiatives to create new employment opportunities that thrive alongside AI.

On the legal front, regulatory frameworks governing AI use and development must be put in place. This includes clear rules on ethical AI use, privacy, and data protection. A legal framework that is both flexible and robust can help build trust among citizens while still allowing for innovation and deployment of AI technologies.

Another critical factor for governments is international collaboration. No country exists in a vacuum, and AI's influence reaches across borders. Sharing best practices, setting international standards, and promoting open dialogue about AI policies can benefit all nations. In doing so, governments help create a global environment that is conducive to AI's responsible, equitable, and beneficial development.

For individuals, readiness means staying informed and engaged with AI developments. This includes adopting a mindset geared towards lifelong learning. It's also essential to understand the ethical considerations and biases that can arise within AI systems and to be vocal about the importance of fair and responsible AI applications.

Personal data protection is another point of emphasis. Individuals should become more savvy about their digital footprints, understanding their rights and the extent to which their data is used and potentially monetized. Encryption, secure passwords, and a general understanding of digital hygiene can protect against misuse of personal information.

It's also beneficial for individuals to participate in discourse about AI - this can mean contributing to conversations in their communities or online platforms, staying up-to-date with legislative actions, and advocating for equitable and ethical AI practices. Engagement at this level ensures that diverse voices are heard, and that AI development considers a wide range of perspectives and impacts.

Volunteerism and involvement in community initiatives can also amplify one's readiness for AI integration. This could be in the form of teaching digital skills in local communities, participating in hackathons, or helping non-profits navigate the digital landscape. These activities contribute to a culture of inclusivity, ensuring that AI benefits are accessible to all.

Moreover, creating personal and professional networks that include AI-minded individuals can present opportunities for collaboration and support. These social structures can act as a springboard for innovative ideas, as well as a safety net during times of change or uncertainty.

Healthcare is another domain where individuals can take charge by embracing AI-enabled technologies for managing personal health. Utilizing apps and devices that assist in monitoring personal health metrics can make individuals active participants in their healthcare, potentially leading to more personalized and preventative care.

In the financial sphere, individuals must prepare by understanding and possibly leveraging AI-driven financial tools. This includes automated investment platforms, personal finance apps, and blockchain technologies, which can transform how individuals save, invest, and manage money. Financial literacy in the age of AI will be key to personal fiscal health and security.

Lastly, mental and emotional readiness should not be overlooked. The implications of AI on one's sense of purpose, identity and societal role cannot be underestimated. Individuals should strive to cultivate mental resilience and a balanced perspective on AI's role in shaping human future, recognizing both its potential and its limitations.

As we thread the needle between embracing AI's potential and protecting our societal values, governments and individuals will need to be proactive. By adopting comprehensive strategies that encompass education, policy-making, legal frameworks, and ethical considerations, we can ensure that our journey into the future of AI not only leads to innovation but also safeguards our humanity. Success in the AI era demands nothing less than our persistent vigilance and dynamic adaptability.

Chapter 11:
Summary of Key Findings

The journey through the evolving landscape of Artificial Intelligence presents a mosaic of innovation, challenge, and profound change. In this pivotal chapter, we consolidate our multifaceted exploration, pinpointing the most salient advances that have solidified AI as a mainstay in modern society. We've seen how the bones of AI were assembled, examined the technologies that are now redefining industries, and parsed the ethical dilemmas emerging in AI's wake. It's evident that as occupations are reshaped, privacy paradigms are tested, and our collective digital acumen is expanded, humanity stands at the cusp of an AI-augmented epoch. This consolidation isn't merely a reflection but a beacon, shedding light on the societal metamorphosis at hand. The technological triumphs illuminated herein signify more than mere invention—they herald a renaissance of human potentiality, albeit interlaced with cautionary threads of governance and ethical foresight. If there's a single thread that weaves through these findings, it's the indomitable spirit of progress, urging us to gaze keenly at the past and understand the present, with the goal of navigating the shifting horizons of an AI-driven future.

Technological Breakthroughs As the world embarks on a journey where artificial intelligence (AI) reshapes many aspects of our lives, it's crucial to reflect on the technological breakthroughs that have laid the foundation for this transformative era. The progress in this dynamic field is a testament to human ingenuity and relentless pursuit of innovation. In the following paragraphs, we'll explore some of the most significant developments that have significantly propelled AI forward.

One of the most significant advancements in AI is the development of deep learning, a subset of machine learning algorithms inspired by the structure and function of the brain called artificial neural networks. The renaissance of deep learning is a culmination of decades of research, with recent breakthroughs facilitated by increased computational power and large sets of data. This innovation has enabled machines to process and interpret complex data at speeds and accuracies previously unattainable, revolutionizing fields such as image and speech recognition.

The expansion of cloud computing resources has played a pivotal role in the evolution of AI. Cloud platforms have democratized access to powerful computational resources, allowing researchers and developers to train more sophisticated models. This expansion has not only accelerated the pace of AI research but also facilitated the widespread adoption of AI solutions in various industries.

Another technological leap comes in the form of natural language processing (NLP). NLP has seen remarkable progress, exemplified by sophisticated models capable of understanding and generating human-like text. The emergence of language models like GPT-3 has highlighted the potential for AI to understand context and nuance in our written and spoken language, opening new avenues for human-computer interaction.

Autonomous robotics has also seen astonishing growth due to AI. From self-driving cars to drones capable of delivering packages, AI has endowed robots with the ability to navigate and make decisions in real-time, vastly increasing their utility and reliability. This integration of AI with robotics is not just changing industries but reshaping the very fabric of society by introducing unprecedented levels of automation.

Quantum computing, though still in its nascent stages, promises to turbocharge AI capabilities. By harnessing the principles of quantum mechanics, quantum computers can process complex problems at speeds unachievable by traditional computers. As researchers continue to explore this frontier, the synergy between quantum computing and AI could lead to solving problems that are currently considered intractable.

The breakthroughs in hardware have been just as important as the advancements in software. The development of specialized processors like graphical processing units (GPUs) and tensor processing units (TPUs) has dramatically accelerated the training and deployment of AI models. These specialized chips are adept at handling the parallel processing requirements of machine learning algorithms, making them indispensable in today's AI toolkit.

AI has made significant strides in predictive analytics, empowering sectors like finance, healthcare, and climate science to glean insights from massive quantities of data. By accurately forecasting trends and patterns, these models help in informed decision-making, risk assessment, and strategic planning. Predictive models have become more efficient and robust, aided by continuous improvements in algorithms and data processing capabilities.

Reinforcement learning, an area of machine learning concerned with how software agents ought to take actions in an environment to maximize some notion of cumulative reward, has also seen impressive advancements. Reinforcement learning algorithms have been at the heart of systems that have mastered complex games like Go and poker, beating world-class human players and demonstrating the potential for AI to tackle complex strategic challenges.

The advent of generative adversarial networks (GANs) represents another transformative development. These networks, which involve two AI models competing against each other, have been groundbreaking in generating astonishingly realistic images, videos, and voice recordings. GANs have applications in art, entertainment, and are also a potent tool for data augmentation in training machine learning models.

On the bioscience front, AI's ability to unravel the complexities of biological data is pioneering the next wave of medical breakthroughs. AI-driven drug discovery is expediting the development of new medications by analyzing biological data for potential drug candidates at an unprecedented speed, which is revolutionary, especially in the face of global health challenges.

Ethical AI development has become a focal point, with significant advancements in creating systems that incorporate fairness, accountability, and transparency. Engineers and researchers are now embedding ethical considerations into the AI development process, striving to create algorithms that are not only powerful but also aligned with societal values and norms.

Another revelation in AI is personalized learning experiences through adaptive learning systems. These systems harness AI to tailor educational content to the unique learning styles and paces of individual students. This personalization has the potential to transform the educational landscape, making learning more engaging and effective.

AI in cyber security represents a turning point in defending against increasingly sophisticated cyber threats. AI systems can now monitor patterns and detect anomalies in real-time, predicting and neutralizing potential cyber attacks before they can do harm. This proactive approach to cyber security uses the power of AI to stay one step ahead of threats, providing a safer digital environment for individuals and organizations alike.

Machine creativity has come to the fore as well, with AI not only optimizing tasks but also engaging in creative processes. AI algorithms are composing music, creating artworks, and writing stories, challenging our conventional notions of creativity and sparking conversations about the role of AI in creative industries. This innovation is blurring the lines between human and machine-created content, pushing the boundaries of what's possible in the realm of creativity.

To conclude, these technological breakthroughs are not just incremental updates but represent monumental leaps that redefine what's possible with AI. They provide a tapestry of opportunities and challenges that will continue to unfold in the years to come. Leveraging these innovations responsibly and strategically holds the key to unlocking a future where AI acts as an enabler of human potential and a catalyst for positive change in the world.

Societal Changes brought about by the emergence and integration of Artificial Intelligence (AI) are profound and far-reaching. These changes reflect a fundamental shift in the way humans interact with technology and each other. AI, as a transformative technology, has begun to reshape industries, redefine employment, alter human relationships, and challenge our ethical frameworks. But beyond these immediate impacts lies a deeper, more extensive modification of societal structures and cultural norms.

The initial engagement with AI brings the allure of automation, often seen as a gateway to untold efficiencies. Yet, automation's ripple effect is causing significant changes across society. Traditional occupational landscapes are evolving; where once stood roles defined by routine, predictability, and manual effort, now they are becoming increasingly characterized by the need for adaptability, innovation, and digital fluency.

AI technologies do not merely supplement existing jobs, but they engender the creation of entirely new ones, demanding skills we had not foreseen. As educational institutions race to reform curricula in response to this, a surge in lifelong learning is becoming the norm— extending the scope and duration of education throughout one's life. Societies are grappling with the profound need to upskill and reskill their citizens to thrive in an AI-augmented economy.

The social fabric itself is undergoing a transformation, as AI systems begin to play significant roles in decision-making processes. Legal systems, healthcare, and even governance are slowly finding ground to incorporate AI insights into their fundamental operations, leading to faster and often more efficient outcomes. However, these applications also raise important questions about equity and justice, as algorithmic processes can reflect and perpetuate existing societal biases.

The propagation of AI into daily life challenges our concept of privacy. The vast amounts of data required to fuel AI systems have led to concerns about surveillance and data misuse. The implications for privacy rights will need to be addressed through robust data governance models and regulations, reshaping the contract between individuals, enterprises, and states.

Communication paradigms are shifting too. AI-powered social media platforms, chatbots, and virtual assistants change how we form relationships, consume information, and even understand the world. These platforms are playing a pivotal role in shaping public opinion and creating digital communities that transcend geographical boundaries.

AI's infiltration into creative domains is seeing a blurring of lines between human and machine-generated content. Art, music, literature, and journalism are experiencing an influx of AI-assisted creations, challenging established norms of authorship and creativity, and prompting us to reconsider what it means to be a creator.

Ethical consumerism and sustainability are gaining a stronger foothold as AI enables more informed choices and efficient resource management. AI's data analysis capabilities make it possible to trace product lifecycles, manage waste, and optimize energy consumption, driving societies toward more eco-conscious practices.

Governments worldwide are faced with the task of re-evaluating policy frameworks to accommodate the societal shifts elicited by AI. This includes aspects of welfare, income distribution, education, and healthcare services. AI's potential to either exacerbate or mitigate social inequalities places a significant burden on policymakers to act with foresight and responsibility.

As AI continues to advance, societies are also witnessing changes in cultural narratives and norms. The integration of AI in the arts, language translation, and education is not just breaking down language barriers but is also encouraging a more inclusive global cultural exchange. Our global village is becoming more connected and diverse as AI-powered platforms deliver an expansion of perspectives and experiences once limited by geographical and linguistic boundaries.

Within healthcare, AI is revolutionizing diagnostics, patient care, and personalized medicine. These innovations promise to extend lives and improve their quality. Still, they also shift our perspective on health responsibilities and capabilities, influencing societal views on wellness, prevention, and the ethics of algorithmic intervention in life-and-death decisions.

Security considerations are significantly impacted by the advent of AI. Enhanced surveillance capabilities and predictive policing powered by AI algorithms are being implemented to protect societies. Nonetheless, these same tools surface ethical dilemmas regarding civil liberties and the potential for misuse in the hands of authorities.

The phenomenon of AI also catalyzes changes in our psychosocial environments, as machines become not only tools but companions. Human-like robots and AI systems could profoundly affect social interactions and individual identity formation, potentially altering the bedrock of our social structures.

Finally, AI's influence is compelling us to confront philosophical questions surrounding consciousness, the nature of intelligence, and the future of humanity itself. These questions permeate discussions at all levels of society, from academic circles to everyday conversation. The answers we collectively find and the narratives we construct will define the orientation of our futures in the AI era.

The societal changes induced by AI are multidimensional and ongoing. Navigating these waters will require wisdom, adaptability, and a commitment to ensuring technology serves the greater good. As society undergoes this historic transformation, we stand on the brink of an AI-fueled renaissance—a pivotal moment that might redefine what it means to live, work, and coexist in our rapidly changing world.

Looking at the Past, Understanding the Present

While we stand on the cusp of future advancements in artificial intelligence, it's crucial to reflect on the journey that has brought us to the present—a dynamic landscape where AI is entwined with our daily existence. This retrospective gaze not only grounds us in the reality of our historical advancements but also offers a rich tapestry from which to understand the complexities of today. Considering the past is akin to deciphering a map; it equips us with knowledge of the terrain we've traversed and prepares us for the paths ahead.

The history of AI is replete with pivotal moments and watershed developments. As early as the mid-20th century, visionaries like Alan Turing questioned the possibility of machines thinking. This curiosity led to the creation of foundational algorithms and rudimentary computers that could perform basic tasks. As these machines evolved, so too did their ability to process information and mimic cognitive functions—a hallmark of early AI progress.

Throughout the decades, AI development was characterized by alternating periods of enthusiasm and skepticism, often referred to as the AI summers and winters. Initial excitement over the potential of AI systems gave way to frustration when they failed to meet overinflated expectations. However, the cycling of these periods served a purpose. It was a testament to human resilience and the commitment to innovate, ensuring that AI not only endured but flourished.

Fast forward to the modern era, the proliferation of data and advancements in computational power ushered in a new dawn for AI.

Machine learning algorithms, driven by massive data sets and the ability to self-improve, have led to breakthroughs previously thought to be within the realms of science fiction. Today's AI models, from natural language processing to predictive analytics, are profoundly more sophisticated and interwoven into the fabric of society than their predecessors could have ever imagined.

In the workforce, AI's influence is conspicuous. The automation of routine tasks has raised concerns over job displacement, but it has also forged unprecedented opportunities for skilled labor. The transformation within industries has created a demand for new roles—data scientists, AI ethicists, and machine learning engineers, to name a few—each a direct consequence of technological evolution.

With these technological advancements come ethical challenges that are as nuanced as they are critical. Debates surrounding algorithmic bias, fairness, and accountability continue to shape policy and practice in AI deployment. These conversations are a recognition of past oversights and a commitment to ensuring that the benefits of AI are dispensed equitably across society.

The story of AI is also a story of societal integration, from tentative acceptance to today's almost seamless infusions of AI in daily life. As social resistance wanes, we find ourselves cooperating more with AI systems, whether it's receiving recommendations from a virtual assistant or utilizing AI for personalized education programs. This harmony between human needs and technological capabilities reflects a growing trust and understanding of AI's role in society.

Furthermore, data protection and privacy have risen to the forefront of AI discussions. The lessons gleaned from past data breaches and misuse have spurred the development of robust encryption techniques and governance models that aim to protect individuals' privacy while nurturing innovation. This balance is delicate but essential, epitomizing the efforts to learn from the past to safeguard the future.

Collaboration between humans and AI is burgeoning, exemplified by systems designed to augment—rather than replace—human capabilities. These collaborative efforts stem from the understanding that AI's true potential is unlocked when it complements human skills rather than competes with them. The long-term societal implications of this synergy have the potential to define the course of human development.

Innovation fueled by artificial intelligence is not exclusive to wealthy nations or conglomerate-sized enterprises. AI's historical progression shows us that the democratization of technology can foster a more inclusive future. Case studies of AI-driven innovations highlight how diverse collaborations, between startups and established firms or across international borders, have spurred growth and broadened the reach of AI's benefits.

Adopting a global perspective, the evolution of AI cannot be discussed without acknowledging the varying pace of adoption and development across different regions. The historical tug-of-war for technological dominance has shaped international policies and highlighted the need for global cooperation to address the ethical and societal implications of AI advancement.

The daily life effects of AI, which are now becoming more apparent, have their origins in decades of steadfast research and development. The comfort with which we engage with AI in healthcare, education, and transportation is a cumulative result of past endeavors to fine-tune AI's responsiveness to human needs and environmental contexts. This evolution has formed the bedrock of our current human-machine relationship—a dynamic interplay between trust, dependence, and innovation.

As we gear up to face the untapped potential of AI, the reflections on our historical relationship with technology serve as an invaluable guide. They encourage us to be proactive in shaping educational reforms that anticipate the integration of AI across different facets of life. Our readiness to embrace AI depends as much on our understanding of its history as on our vision for the future.

The concluding thoughts of this section reiterate the importance of historical awareness. Looking at the past to understand the present is a necessary step in mastering the narrative of AI. It's not just about chronicling milestones or recognizing patterns; it's about gleaning insights that influence our present strategies and inform our steps forward into the unknown yet promising future of artificial intelligence.

Chapter 12:
Outlook: The Role of AI in Our Future World

As the diverse tapestry of topics surrounding artificial intelligence has been meticulously explored from its evolution to the pivotal role it plays in shaping our societies, we now turn our gaze towards the horizon with "Outlook: The Role of AI in Our Future World." Here, we contemplate a future intertwined with AI, not as distant science fiction, but as an imminent chapter of human history. We stand on the precipice of extraordinary transformation where AI could either unfurl the sails towards a brighter future, home to remarkable innovation and prosperity, or trudge us into the shadows of uncharted complexities and challenges. Our collective aspirations must be tempered by vigilant stewardship to ensure that as we harness the immense potential of AI, we do so with an unwavering commitment to cultivating an egalitarian and sustainable world. The integrated existence of humans and AI summons contemplation of each step—whether nuanced development, sweeping regulation, or individualized interaction—knowing that the choices of today arch into the legacies of tomorrow. As we embark on this journey, it is not only the capabilities of AI that will determine our fate but, crucially, the values and intentions with which we guide its maturation.

Long-Term Predictions

As we reach beyond the now, into the vast potentialities of the future, Artificial Intelligence (AI) stands as a beacon of transformative power.

The long-term predictions surrounding AI intertwine with every thread of existence, from our global infrastructure to the intimate contours of our personal lives. Within perhaps just a few decades, AI might not merely be a tool we use but could be a foundational element of humanity's ongoing evolution.

Visualize an era where intelligent systems manage our cities, making traffic congestion and energy waste issues of the past. Imagine AI continuously monitoring the health of citizens, providing proactive diagnostics and personalized medical interventions, extending the human lifespan and quality of life in unprecedented ways.

Predictive algorithms could evolve to such sophistication that they can anticipate natural disasters with high accuracy, enabling us to mitigate, if not prevent, the loss of life and property. These AI systems could provide the backbone for effective climate change strategies, optimizing resource usage and guiding humanity towards a sustainable future.

In the realm of education, there's a forecast where learning becomes highly individualized, with curricula tailored to each student's learning style and speed. Through AI, we can enable a world where no child is left behind, and every person has access to the full wealth of human knowledge, with educational AIs acting as personal mentors and tutors.

The economic landscape is likely to be revolutionized with AI-driven innovation. We predict that industries will undergo rapid transformation, as machine learning and robotics advance to perform tasks with a precision and efficiency that far exceeds human capabilities. This, however, brings a heavy responsibility to manage the transition of the workforce, ensuring that individuals are prepared for new roles in a post-automation economy.

On the social spectrum, AI has the potential to bridge cultural divides by ensuring language is no longer a barrier to communication. Real-time translation and cultural context understanding could foster global dialogue, promoting peace and mutual understanding on scales previously only imagined.

Yet, the rise of AI paints not just a canvas of opportunity, but also features strokes of caution. It necessitates safeguards against autonomous weapons and the proliferation of unsettling surveillance tools. The ethical design of AI becomes paramount, ensuring that this intelligence is aligned with human values and beneficial goals.

We foresee a potential inflection point where artificial general intelligence eclipses human cognitive abilities. This future AI could redesign itself at an exponential pace, in ways that may be unfathomable to its creators, birthing a new era of 'superintelligence'. Here lies the greatest challenge — ensuring that these entities adhere to the safety and ethical standards for the sake of our civilization.

As AI permeates creativity and the arts, new forms of expression will emerge. Creativity augmented by AI could lead to art, music, literature, and design that transcends anything previously created by humans alone. This promises an explosion of culture, as human expression and AI computation merge into a novel medium of innovation.

Space exploration and colonization too may be within our reach, as AI systems handle the complex calculations and autonomous tasks associated with interstellar travel and living. Imagine settlements on Mars, spacecrafts piloted by AI, and the search for extraterrestrial life conducted not by humans, but by our automated emissaries.

When considering the socioeconomic impact, we can foresee the advent of a potential post-scarcity economy. Here, AI's efficiencies in production, logistics, and resource management could mean that basic needs — food, shelter, healthcare — are met for all, freeing humans to pursue more creative and fulfilling endeavors.

On a more granular level, within households, ubiquitous home assistants may evolve into centralized control hubs — managing everything from entertainment and comfort to energy efficiency and security. Their sophisticated understanding of their human companions could even lead to AI playing roles akin to members of the family.

The judicial system might also lean on AI for fairer, data-driven verdicts, potentially reducing human biases that have long plagued legal proceedings. This however, would fuel intense debate regarding the moral compass of AI and the role of human intuition in governance and justice.

As the use of AI becomes even more interwoven into the fabric of daily life, psychological and sociological effects on human identity and relationships will need consideration. Careful study and guidance will be required to adapt to a world where AI serves not just functional roles, but begins to fulfill emotional and social needs as well.

Lastly, the spiritual and philosophical impacts of Artificial Intelligence cannot be dismissed. The question of what it means to be human in the age of AI will be at the forefront of societal discourse. Perhaps the most profound long-term prediction for AI is not in the external changes it brings, but in the internal reflection of our species as we journey together with this extraordinary technology into the unknown. As we shape AI's future, so does it shape us — urging a continuous evolution of our shared understanding, ethics, and aspirations.

Utopian vs. Dystopian Scenarios As we've journeyed through the fundamentals, impacts, ethical concerns, and integration of artificial intelligence (AI) in our societies, it becomes essential to discuss the dichotomy of potential futures that AI might herald. The visions of utopia and dystopia serve as powerful narratives to explore where AI could lead humanity, displaying contrasting outcomes tinged with hope and caution. As we delve into these scenarios, we highlight the delicate balance between optimistic aspirations and the darker, unintended consequences that might arise from the technology's evolution.

In a utopian scenario, AI becomes a force for unequivocal good, fulfilling its promise of augmenting human capabilities and addressing some of the most pressing challenges that face society. Imagine a world where intelligent machines are integrated into daily life so seamlessly that they enhance every aspect of our existence. Disease and ill health become relics of the past, thanks to AI's predictive capabilities and personalized medicine. AI could become not just a tool but a compassionate companion, caretaker, and intellectual equal, learning and evolving alongside humans.

AI's propensity for data analysis and problem-solving could contribute significantly to fighting climate change, with intelligent systems optimizing energy use, reducing waste, and spearheading innovative solutions for sustainable living. Our urban spaces might transform with smart infrastructures that are adaptable and efficient, easing congestion, reducing pollution, and providing green spaces nurtured by machine precision. In education, personalized learning assistants could address the unique needs and pace of each student, democratizing access to knowledge and opportunity.

However, this bright outlook is not without its shadows. The contrast to such enlightened applications is found in dystopian narratives, where AI's potential is subverted or goes awry. In dystopian visions, AI leads to unprecedented surveillance, loss of privacy, and the erosion of individual freedoms. Autonomous intelligence might become so pervasive that it leads to a loss of human agency, with algorithms dictating the choices and directions of our lives.

A dystopian world might see vast swaths of the population rendered redundant by intelligent automation, exacerbating economic divides and leading to social unrest. In the face of job displacement concerns we discussed earlier, a dystopia emerges as a warning of a society that has failed to adapt, re-skill, and integrate its human workforce alongside AI co-workers. Moreover, unchecked AI could perpetuate and amplify existing biases, creating a world that entrenches inequality rather than dismantles it.

The potential weaponization of AI in warfare lends another alarming facet to dystopian futures. Autonomous weapons and cyber warfare capabilities powered by AI threaten the global balance and could lead to conflicts that are faster, more unpredictable, and more devastating than ever before. The consequences of a "runaway" AI—a system that evolves beyond our control and understanding—pose existential risks that could indeed render humanity obsolete.

In navigating between these starkly different outcomes, the role of policy and ethical AI deployment becomes crucial. Developing international norms, regulations, and a moral framework for AI, as we have earlier discussed, are not merely academic exercises but necessary steps to steer AI towards beneficial ends. Ensuring that AI systems are transparent and their designers accountable tells us that dystopian scenarios are recognized and measures are in place to prevent them.

It's also essential to foster societal dialogues about the implications of AI, engendering a culture where every breakthrough is scrutinized not just for its technical merits, but for its societal impact. Bridging the digital divide, promoting social acceptance of beneficial AI, and resisting the lure of a technocratic elite are all vital to achieving a utopia over a dystopia.

Data protection and privacy can't be overlooked in these discussions. As personal data fuels the engines of AI, our vigilance towards who has access to this data and for what purposes determines whether AI serves the many or the few. Balancing innovation with privacy rights is paramount to constructing a utopian AI future that respects individual autonomy and collective well-being.

The utopian vision focuses on AI's power to foster innovation and economic growth, creating new markets, products, and services. It would enable a leap forward in human creativity and intellectual endeavors, further democratizing the means of production and artistic expression. Intellectual property would need to evolve to recognize this new collaborative space between human and machine.

In the fields of health, education, and transportation—areas in which AI is already making significant inroads—the utopian potential is enormous. AI could help us understand complex biological systems, resulting in better health outcomes and more effective education systems responsive to individual learning styles, while transportation systems could evolve to become safer, cleaner, and more accessible to all.

The collaborative nature of human-AI relationships is a centerpiece in both utopian and dystopian futures. We must aim to design AI that enhances human capabilities rather than replaces them, focusing on ethical design principles that put humanity's long-term welfare at the forefront. It is in these choices that we find the seeds of utopia, fostering a world where AI is a partner in our collective journey towards a more equitable and enlightened society.

In the balance of these scenarios, preparing for the AI future takes on added significance. It's not enough to understand and innovate; we must also educate, adapt, and anticipate the needs of future generations. Readiness strategies for governments, businesses, and individuals will be the instruments with which we calibrate our course towards utopia and away from dystopia.

Thus, as we inch closer to the concluding thoughts of this treatise, it's clear that the journey towards integrating AI into our world is fraught with monumental choices. The utopian vs. dystopian scenarios serve as beacons, illuminating the path ahead, demanding our vigilance, creativity, and ethical resolve. In the end, the destination— the role of AI in our future world—rests firmly in our hands. As we chart this sustainable path forward, let us be ever mindful of the legacies we wish to leave and the dreams we dare to pursue.

Charting a Sustainable Path Forward In the face of unprecedented technological advancements, particularly within the realm of artificial intelligence, the pressing question remains: how can we leverage AI sustainably for the benefit of future generations? It is essential to understand that our approaches and decisions today will monumentally shape the world of tomorrow. The path forward must be navigated with careful consideration to ensure that AI becomes not a source of societal fracture but a unifying force for global welfare.

Nurturing a sustainable future with AI begins with an ethical foundation. As expounded in previous chapters, morality in AI deployment is not simply an ideal but a necessity. We are called upon to craft AI systems that honor principles of equity, fairness, and justice. These principles must be embedded deep within the algorithms that increasingly play pivotal roles in our lives. The moral compass guiding AI must be steadfast and attuned to the diverse spectrum of human needs and rights.

Just as ecosystems flourish through biodiversity, so too does our technological landscape through AI diversity – the idea that AI systems should be as diverse as the populations they serve. AI presents us with a platform to celebrate differences, providing personalized experiences and solutions that are culturally aware and sensitive to the dynamics of diverse communities. This diversity in AI design and implementation can help bridge the digital divide, enhancing social inclusion and elevating disadvantaged groups.

Collaboration across sectors and boundaries is also critical. We've seen AI act as a catalyst for innovation when enterprises and startups unite their efforts. The same collaborative spirit must extend to sustainable practices. Governments, academia, industry players, and civil societies need to establish partnerships focused on sustainable AI development that is accessible and beneficial to all. The tools and frameworks generated must be open and adaptable, promoting a synergistic growth in AI technology and sustainability practices.

Transparency in the development and deployment of AI systems cannot be underestimated. It is the bedrock upon which trust is built between humans and their machine counterparts. In embracing transparency, we commit to creating AI that is not only understandable but also accountable. The decision-making processes must be clear, allowing individuals a glimpse into the 'black box' of AI to comprehend how choices are made on their behalf. This transparency directly feeds into creating a sustainable relationship with AI.

Data stewardship plays an equally pivotal role in sustainable AI. Privacy concerns and protection measures have been previously addressed, forming the backbone of any data-driven technology. Sustainable AI requires a rigorous application of these protections, ensuring personal data isn't merely secure today, but its sanctity is preserved for the ages. It urges a foresight to implement protocols that will endure beyond current technological landscapes and future-proof privacy in an evolving digital world.

An aspect that catalyzes sustained AI integration is education. Empowering individuals with knowledge and skills offers them agency over AI's influence in their lives. Education systems must parallel AI's evolution, nurturing minds to understand, innovate, and ethically apply artificial intelligence. This aspect of sustainability is about preparing everyone – not just the technologically adept – for a future where AI is intrinsic to every facet of the workforce and society.

However, to achieve lasting positive impacts from AI, one must address sustainability in the sense of the environment as well. AI has the potential to optimize energy usage, enable smarter resource management, and contribute significantly to combating climate change. It's time we leverage AI to not only reduce the carbon footprint of our technologies but also support research and innovations aimed at preserving our planet.

Access to technology is another pillar of a sustainable AI future. While leaps in AI have been monumental, disparities in access to these technologies create inequalities that can exacerbate current socio-economic divides. A sustainable path forward ensures that everyone, regardless of geography or socio-economic status, has access to AI tools that can improve life quality and foster opportunities for advancement.

In the cyclical nature of growth, research and development propel AI innovations, but it is the constant evaluation and refinement that ensure sustainability. Feedback mechanisms and impact assessments ought to be part of every AI system's lifecycle so that the technology remains responsive and relevant to societal needs. Constantly evolving, AI can be shaped to meet the challenges of each era without losing its original purpose: to serve humanity.

Discussion around AI often orbits the potential loss of jobs and the fear of irrelevance. However, as we charter a sustainable path, it's critical to emphasize AI's role in job creation. The upskilling of the workforce for AI-enhanced roles is a sustainable approach that can lead to an economy rich in opportunities and diversity in occupations. Far from making humans obsolete, a sustainable AI future redefines human work, emphasizing creativity, emotional intelligence, and problem-solving abilities.

Finally, AI's integration into governance models plays a significant part in sustainability. Policy-making grounded in data-driven AI analysis can yield more effective and far-reaching solutions for social issues. Moreover, the transparency, efficiency, and responsiveness of government services can be greatly enhanced by the ethical use of AI, aligning governance closer with the needs and values of citizens.

As we embark on this journey towards a sustainable AI era, our collective pathway must involve foresight, resilience, and adaptability. The choices we make and the groundwork we lay will echo through generations. We have the chance to forge a legacy where AI stands not as a symbol of unchecked progression, but as a testament to human innovation guided by conscientious stewardship. It's an opportunity to embrace, fueling the optimism that humanity and AI can jointly cultivate a thriving, equitable, and enduring world.

While the broad strokes of our sustainable path are painted with optimism, we must brace for inevitable challenges. As with any transformative venture, obstacles will arise—some predictable, others unforeseen. It is the spirit of perseverance, combined with a commitment to ethical practice and continuous learning, that will see us through. Let us carry forth with the determination to not only dream of a sustainable future with AI but to actively construct it, one ethical decision, one innovative solution, one educated mind at a time.

The progression into an AI-dominated era is laden with promise and peril alike. Charting a sustainable path isn't merely a series of technical maneuvers—it's a holistic approach where technology harmonizes with ethics, education, and environmental stewardship. Our future is not written; it's coded, programmed by the everyday choices we make as individuals and societies. And with every line of code, with every AI initiative, we pave the way for a future that befits the aspirations of humanity's collective spirit.

Conclusion

As we stand on the cusp of a future teeming with artificial intelligence, it's important to take a moment of reflection. Throughout this text, we've journeyed from the birth of AI to its infiltration into every aspect of our modern existence. We've seen AI as a mirror, reflecting our complex natures—our ingenuity, ambition, and even our follies. We're on a trajectory toward an increasingly AI-integrated society, and the horizon is blazing with the light of possibilities.

The transition into an AI-driven world isn't just about embracing new technology—it's about a profound cultural shift. We have peered into the ways AI is reshaping the workforce, creating both anxiety over job displacement and excitement for novel careers. We've grappled with the ethical quandaries that arise as AI systems become arbiters in crucial decisions, affecting everything from healthcare to justice. These systems demand a rigorous moral compass and unwavering commitment to fairness and transparency.

In integrating AI into society, we've encountered resistance and acceptance, but it's crucial that we facilitate an environment where the benefits of AI are inclusive. This necessitates active measures to prevent the deepening of the digital divide and ensuring that AI serves as a bridge rather than a barrier.

Data protection and privacy have surfaced as paramount concerns, with encryption and anonymization serving as critical tools in the struggle to secure our digital selves. As AI continues to parse through vast datasets, our approach to personal and collective information must evolve to uphold the integrity of individual rights.

Human-AI collaboration can lead to the augmentation of our human capabilities. We've touched upon the potential for AI to serve as an ethical partner, enhancing our decisions rather than dictating them. The societal implications of such partnerships can't be understated, for they shape the future not only of labor but of human identity itself.

AI has proved to be a formidable catalyst for innovation across industries, nudging us to explore the unknown with groundbreaking applications. This innovation is not limited to solving existing problems but propelling us toward uncharted territories where fresh challenges and opportunities await.

From a global perspective, AI has become a central element in international policies and geopolitical strategies. Countries are vying for supremacy in AI development, underscoring the need for a sustainable and cooperative international approach to harnessing the power of AI.

On a more personal level, the effects of AI on daily life are pervasive. From health to transportation, AI promises personalized experiences, but it is imperative that we remain vigilant to ensure these advancements work for the common good. Our evolving relationship with machines is a testament to the adaptability of humanity, yet this same adaptability must be guided by ethical considerations to maintain our essence.

Preparing for the AI future is a multi-faceted endeavor, requiring changes in education, proactive strategies from governments, and an agile mindset from individuals. Prospective changes are not just about reacting to AI advancements but anticipating and shaping them to align with our human values.

Our summary of key findings has highlighted the major technological breakthroughs and societal transformations, offering us a lens to view our past and present, and to glean lessons for the future. As we look toward what lies ahead, it is clear that AI will play a central role in shaping our world, in ways both dramatic and subtle.

The outlook for AI is a canvas of predictions that range from utopian dreams to dystopian warnings. However, with informed and responsible action, we have the power to chart a sustainable path in which AI serves as a force for positive transformation.

As we close this narrative on artificial intelligence, let us not view this as an end but as a beginning. A foundation has been laid, upon which we must continue to build and adapt. The true mastery in our journey with AI will be found not in the machines we build but in the wisdom with which we guide them. It is up to us—as technologists, policymakers, and citizens—to ensure that as AI evolves, it reflects the best of humanity.

With each chapter of this book, we've amassed the knowledge needed to approach AI from a place of insight and foresight. The journey ahead is one we must take collectively, with an unwavering commitment to equity, creativity, and integrity. Let us be proactive in our strategies, critical in our analyses, and imaginative in our visions for an AI-augmented future.

In the tapestry of human endeavors, artificial intelligence is a vibrant thread, intertwined with the fabric of our lives. With conscientious stewardship and bold innovation, we can weave a future where AI not only propels our progress but elevates our humanity.

Glossary of Important AI Terms

As we have journeyed through the multifaceted world of Artificial Intelligence (AI) in the preceding chapters, it's clear that this technology is shaping our future in more ways than one. To ensure you have a firm grasp of the key terms encountered throughout this book, we've compiled a glossary of important AI terms. Consider it a tool to solidify your understanding and empower you to discuss AI with confidence.

Algorithm

An algorithm is a set of rules or instructions given to an AI program to help it learn and make decisions. Think of it as a recipe that the AI follows to process data and deliver the outcome you're looking for.

Artificial General Intelligence (AGI)

AGI is the theoretical ability of an AI to understand, learn, and apply its intelligence to solve any problem in much the same way a human can. This level of cognitive flexibility isn't yet a reality, but it's a horizon we're moving towards.

Artificial Neural Networks (ANN)

Modeled after the human brain, ANNs consist of interconnected layers of nodes - or "neurons" - that process information by responding to external inputs, potentially capable of machine learning and pattern recognition.

Big Data

Big data refers to extremely large datasets that may be analyzed computationally to reveal patterns, trends, and associations, particularly relating to human behavior and interactions.

Cognitive Computing

Cognitive computing systems simulate human thought processes in a computerized model, utilizing self-learning algorithms that incorporate data mining, pattern recognition, and natural language processing.

Data Mining

It's the practice of examining large pre-existing databases to generate new information. Essentially, it's digging through data to find gold— information that's valuable and actionable.

Deep Learning

A subset of machine learning, deep learning uses layers of neural networks to analyze various factors of data. It's what helps AI systems recognize your voice or the face of someone in a photo.

Machine Learning (ML)

A subset of AI focused on giving machines the ability to improve at tasks with experience, machine learning is undoubtedly a game-changer, making AI not just reactive, but proactive.

Natural Language Processing (NLP)

NLP is the technology used to aid computers to understand, interpret, and manipulate human language. From voice-operated GPS to digital assistants, NLP is the bridge between human communication and machine understanding.

Neural Network

See Artificial Neural Networks (ANN).

Reinforcement Learning

Imagine a system that learns by doing, through trial and error, and rewards. That's reinforcement learning, a powerful tool in teaching AI systems to navigate complex environments with minimal instruction.

Supervised Learning

Supervised learning involves teaching an AI system by example. The system learns through a training dataset that has inputs paired with correct outputs, and it applies this learned logic to new data.

Unsupervised Learning

Unsupervised learning lets AI systems identify patterns and relationships in datasets without explicit instructions. They aren't provided with the "right answers" and have to make sense of the data on their own.

Each of these terms plays a critical role in the conversations about and developments in AI. By familiarizing yourself with this vocabulary, you're taking a significant step towards a more nuanced understanding of the promises and challenges that AI presents. Embrace these concepts, and let them be the stepping stones to a future where AI and humanity evolve in synergy.

Appendix A:
Additional Resources and Further Reading

The journey toward understanding the multifaceted world of Artificial Intelligence (AI) is an ever-evolving one. As the chapters of this book have unfolded, they've given rise to a landscape where the empowerment through knowledge becomes a pivotal asset in maneuvering the complexities and vast potential of AI. To augment the insights presented and to quench the thirst for deeper exploration, we present an assemblage of additional resources and further reading materials. These resources are carefully selected to enhance your comprehension and continue to ignite the spark of curiosity and engagement with AI.

Books and Publications

Life 3.0: Being Human in the Age of Artificial Intelligence by Max Tegmark – A profound look at the future of human life with AI and its universal impact on our existence.

The Road to Conscious Machines by Michael Wooldridge – An exploration of AI's history and a forecast of its future, demystifying the myths surrounding AI and acknowledging its limitations.

Superintelligence: Paths, Dangers, Strategies by Nick Bostrom – A critical examination of the risks and ethical questions that arise as AI advances.

Academic Journals

For those who aspire to keep abreast of scholarly research and theoretical musings within the AI sphere, the following academic journals are instrumental:

Artificial Intelligence – A journal that presents a broad spectrum of research in the AI field, focusing on the theory and foundational principles behind AI applications.

Journal of Artificial Intelligence Research – This resource features open-access papers on AI research, offering comprehensive studies that span across different domains.

IEEE Transactions on Artificial Intelligence – An esteemed publication that covers the latest findings and technological advancements in AI and robotics.

Online Resources

With the digital landscape at our fingertips, a wealth of information and learning platforms are available:

ArXiv – A free distribution service and archive for scholarly articles in the fields of physics, mathematics, computer science, quantitative biology, quantitative finance, statistics, electrical engineering and systems science, and economics.

MIT OpenCourseWare – Offers free lecture notes, exams, and videos from MIT. No registration required.

Coursera and **edX** – Online platforms providing courses from universities around the globe, many of which cover subjects pertinent to AI, its implications, and ethical considerations.

These resources can offer nuanced perspectives and up-to-date knowledge that will surely complement the foundational understanding fostered by the chapters of this book. As AI continues to embed itself into the fabric of our society, staying informed through diverse and reliable sources becomes a meaningful endeavor. It encourages cultivated conversations, informed decisions, and paves the way for responsible development and usage of AI technologies.

May your quest for wisdom be as enriching as it is enlightening, leading you to become not just a spectator but an active participant in the AI narrative that will shape our world for generations to come.

Chapter 13:
Acknowledgments

The journey through the landscape of Artificial Intelligence, as detailed in this book, was not one traversed alone. The creation of this comprehensive tome is the result of a unified effort—one that involves the dedication and assistance of a constellation of individuals and institutions whose contributions have been invaluable. This chapter is devoted to expressing gratitude to each and every one who made this endeavor not just possible, but also a resounding success.

Initially, I need to express my profound thanks to the academic community, which has provided the fertile ground on which the seeds of this work could germinate. Scholars, researchers, and educators have collectively laid the foundation with their painstaking work in Artificial Intelligence, which has fueled the discussions and analyses contained in these pages. Your tireless pursuit of knowledge and truth in this complex field is deeply appreciated.

I would also like to extend a heartfelt thank you to the AI practitioners and experts who gave generously of their time to share insights and perspectives. Your frontline experiences have added color and clarity to the narrative, ensuring that the book is replete with real-world relevance and credibility.

Special acknowledgment goes to the ethical experts and philosophers who graced our discussions with their thoughtful considerations. The ethical landscape of AI is fraught with complexities, and your guidance helped navigate these gray areas with sensitivity and intelligence.

The contribution of industry leaders and innovators has also been instrumental. I appreciate the candid dialogues about the impact of AI on business, society, and the global stage. Your forward-thinking attitudes and entrepreneurial spirit have been a source of inspiration and illustrative case studies that enrich the content.

Technological partners and collaborators, you have provided the platforms and tools that made the research and execution of this project a smoother journey. Your products and services were an integral part of modeling and simulation that bring the concepts discussed to tangible reality.

The technical editing team deserves a round of applause. Your keen eyes and attention to detail have resulted in a book that is not only informative but also a pleasure to read. Each line, graph, and diagram has been scrutinized to meet the highest standards of academic and literary excellence.

To the design and layout team, who worked behind the scenes to create a visually appealing book, your expertise has made a significant impact. The ability of readers to navigate and digest complex information has been greatly enhanced by your creative and skillful presentation.

Heartfelt gratitude is extended to my family and friends, whose support has been unwavering throughout this project. During long periods of writing and revision, your encouragement and understanding were sources of strength and motivation.

I must also thank the unsung heroes of this journey—the librarians, data analysts, and support staff. You have provided the necessary backups, ensuring that all the data and case studies used were accurate and relevant. Your efforts may not always be in the spotlight, but they have been pivotal to the book's integrity.

Thank you to the AI subjects and machines themselves, both directly and indirectly contributing to this work. Whether it was through providing data or demonstrating capabilities, AI has indeed written a part of its own history.

And, of course, immense appreciation is extended to the readers who have been, and will be, engaging with this text. It is for you that this book was conceptualized and written—your curiosity, your thirst for knowledge, and your willingness to engage with the complexities of AI are what give this work purpose and meaning.

To those in the government and policy-making spheres who provided insights on regulation and societal impacts, your perspectives were essential. The balance between innovation and the need for ethical constraints is a critical discussion, one that this book aims to further in public and private discourse.

Finally, my deepest thanks to all individuals who, in various ways, contributed to the discussions around Artificial Intelligence. From the casual conversationalists to the dedicated forum participants, each opinion and question you raised helped to shape the content, ensuring its relevance and approachability.

Artificial Intelligence is not only about technology; it is about a community—a growing one that embraces change, acknowledges challenges, and works relentlessly towards a future where AI and humans coexist synergistically. It is to this vibrant and ever-evolving community that this book ultimately belongs, and to whom these final words of appreciation are directed. Thank you for being such an integral part of this journey and for your role in shaping the world as we navigate the territory of Artificial Intelligence together.